MCAT®

High-Yield Science: Solutions and Extra Practice

FOURTH EDITION

Edited by Alexander Stone Macnow, MD

KAPLAN

TEST PREP

© 2017 by Kaplan, Inc.

Published by Kaplan Publishing, a division of Kaplan, Inc.
750 Third Avenue
New York, NY 10017

ISBN: 978-1-5062-3153-2

10 9 8 7 6 5 4 3 2 1

Using This Book

This book, the *High-Yield Science: Solutions and Extra Practice* book, is the companion to a core component of your MCAT course, the set of High-Yield Science problems. You will find these High-Yield Science problems in Unit I of your main lesson book. In each science-focused class session of your course, you'll learn MCAT science from one or more of these High-Yield Science problems. In the class session, you'll engage with your lesson book by answering questions and interacting with diagrams. Then, after that class session, you will turn to this book, the *Solutions and Extra Practice* book, to review.

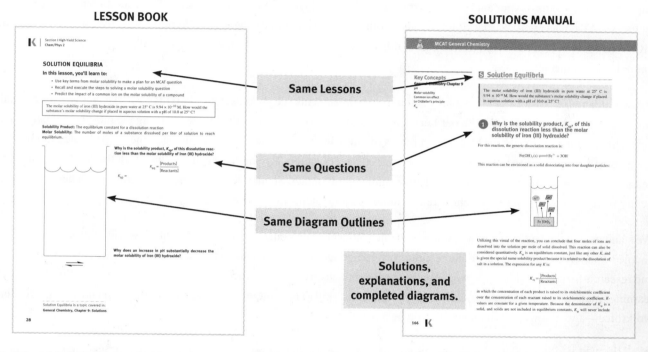

Studying this *Solutions and Extra Practice* book after each of your core class sessions will reinforce the lessons you learn in class. And many of the high-yield science concepts found in this book are bound to be tested on your MCAT.

This book contains worked solutions to High-Yield Science problems. In addition to these High-Yield Science solutions, this book also contains Extra Practice. To help you easily differentiate between the two, High-Yield Science solutions and Extra Practice will be designated using these badges:

"**S**" for High-Yield Science **S**olution   "**E**" for **E**xtra Practice

The best approach is to wait until after you've attended the appropriate class session to study the High-Yield Science solutions **S**. But you can feel free to engage the Extra Practice **E** at any time, including right now! Each chapter of this book progresses through an MCAT science discipline in the same order you'll experience that science in your class sessions. So pick any chapter and start from the beginning of that chapter. It never hurts to work ahead in the Extra Practice **E**. Just make sure to save each High-Yield Science solution **S** until after you've seen that High-Yield Science problem in your class.

Preface

And now it starts: your long, yet fruitful journey toward wearing a white coat. Proudly wearing that white coat, though, is hopefully only part of your motivation. You are reading this book because you want to be a healer.

If you're serious about going to medical school, then you are likely already familiar with the importance of the MCAT in medical school admissions. While the holistic review process puts additional weight on your experiences, extracurricular activities, and personal attributes, the fact remains: along with your GPA, your MCAT score remains one of the two most important components of your application portfolio—at least early in the admissions process. Each additional point you score on the MCAT pushes you in front of thousands of other students and makes you an even more attractive applicant. But the MCAT is not simply an obstacle to overcome; it is an opportunity to show schools that you will be a strong student and a future leader in medicine.

We at Kaplan take our jobs very seriously and aim to help students see success not only on the MCAT, but as future physicians. Our team meets regularly with our Learning Sciences departments to ensure that we're using the most up-to-date teaching techniques in our resources. Multiple members of our team hold advanced degrees in medicine or associated biomedical sciences and are committed to the highest level of medical education. Kaplan has been working with the MCAT for over 50 years and our commitment to pre-med students is unflagging; in fact, Stanley Kaplan created this company when he had difficulty being accepted to medical school due to unfair quota systems that existed at the time.

We stand now at the beginning of a new era in medical education. As citizens of this 21st-century world of healthcare, we are charged with creating a patient-oriented, culturally competent, cost-conscious, universally available, technically advanced, and research-focused healthcare system, run by compassionate providers. Suffice it to say, this is no easy task. Problem-based learning, integrated curricula, and classes in interpersonal skills are some of the responses to this demand for an excellent workforce—a workforce of which you'll soon be a part.

We're thrilled that you've chosen us to help you on this journey. Please reach out to us to share your challenges, concerns, and successes. Together, we will shape the future of medicine in the United States and abroad; we look forward to helping you become the doctor you deserve to be.

Good luck!

Alexander Stone Macnow, MD
Editor-in-Chief
Department of Pathology and Laboratory Medicine
Hospital of the University of Pennsylvania

BA, Musicology—Boston University, 2008
MD—Perelman School of Medicine at the University of Pennsylvania, 2013

Table of Contents

S Solutions

E Extra Practice

See page iii (Using this book) for details

S Solutions

E Extra Practice

See page iii (Using this book) for details

[S] Solutions

[E] Extra Practice

See page iii (Using this book) for details

The *Kaplan MCAT Review* Team

Alexander Stone Macnow, MD
Editor-in-Chief

Tyra Hall-Pogar, PhD
Editor

Bela Starkman, PhD
Editor

Tyler Fara
MCAT Content Manager

Elizabeth Flagge
MCAT Content Manager

Laura L. Ambler
Kaplan MCAT Faculty

Uneeb Qureshi
Kaplan MCAT Faculty

Alisha Maureen Crowley
Kaplan MCAT Faculty

Derek Rusnak, MA
Kaplan MCAT Faculty

Kelly Kyker-Snowman, MS
Kaplan MCAT Faculty

Kristen L. Russell, ME
Kaplan MCAT Faculty

Jason R. Pfleiger
Kaplan MCAT Faculty

Pamela Willingham, MSW
Kaplan MCAT Faculty

Alexandra Côté
Kaplan MCAT Faculty

MCAT faculty reviewers Elmar R. Aliyev; James Burns; Jonathan Cornfield; Nikolai Dorofeev, MD; Raef Ali Fadel; Samer T. Ismail; Elizabeth A. Kudlaty; John P. Mahon; Matthew A. Meier; Nainika Nanda; Caroline Nkemdilim Opene; Kaitlyn E. Prenger; Nicholas M. White; Allison Ann Wilkes, MS; and Tony Yu

Thanks to Kim Bowers; Tim Eich; Owen Farcy; Dan Frey; Robin Garmise; Rita Garthaffner; Joanna Graham; Adam Grey; Allison Harm; Beth Hoffberg; Aaron Lemon-Strauss; Keith Lubeley; Diane McGarvey; Petros Minasi; John Polstein; Deeangelee Pooran-Kublall, MD, MPH; Rochelle Rothstein, MD; Larry Rudman; Sylvia Tidwell Scheuring; Carly Schnur; Karin Tucker; Lee Weiss; and the countless others who made this project possible.

About the MCAT

ANATOMY OF THE MCAT

Here is a general overview of the structure of Test Day:

Section	Number of Questions	Time Allotted
Examinee Agreement		8 minutes
Tutorial (optional)		10 minutes
Chemical and Physical Foundations of Biological Systems	59	95 minutes
Break (optional)		10 minutes
Critical Analysis and Reasoning Skills (CARS)	53	90 minutes
Lunch Break (optional)		30 minutes
Biological and Biochemical Foundations of Living Systems	59	95 minutes
Break (optional)		10 minutes
Psychological, Social, and Biological Foundations of Behavior	59	95 minutes
Void Question		5 minutes
Satisfaction Survey (optional)		5 minutes

The structure of the four sections of the MCAT is further detailed below.

Chemical and Physical Foundations of Biological Systems	
Time	95 minutes
Format	• 59 questions • 10 passages • 44 questions are passage-based, and 15 are discrete (stand-alone) questions. • Score between 118 and 132
What It Tests	• Biochemistry: 25% • Biology: 5% • General Chemistry: 30% • Organic Chemistry: 15% • Physics: 25%

Critical Analysis and Reasoning Skills (CARS)	
Time	90 minutes
Format	• 53 questions • 9 passages • All questions are passage-based. There are no discrete (stand-alone) questions. • Score between 118 and 132
What It Tests	Disciplines: • Humanities: 50% • Social Sciences: 50% Skills: • *Foundations of Comprehension*: 30% • *Reasoning Within the Text*: 30% • *Reasoning Beyond the Text*: 40%

Biological and Biochemical Foundations of Living Systems	
Time	95 minutes
Format	• 59 questions • 10 passages • 44 questions are passage-based, and 15 are discrete (stand-alone) questions. • Score between 118 and 132
What It Tests	• Biochemistry: 25% • Biology: 65% • General Chemistry: 5% • Organic Chemistry: 5%
Psychological, Social, and Biological Foundations of Behavior	
Time	95 minutes
Format	• 59 questions • 10 passages • 44 questions are passage-based, and 15 are discrete (stand-alone) questions. • Score between 118 and 132
What It Tests	• Biology: 5% • Psychology: 65% • Sociology: 30%
Total	
Testing Time	375 minutes (6 hours, 15 minutes)
Total Seat Time	453 minutes (7 hours, 33 minutes)
Questions	230
Score	472 to 528

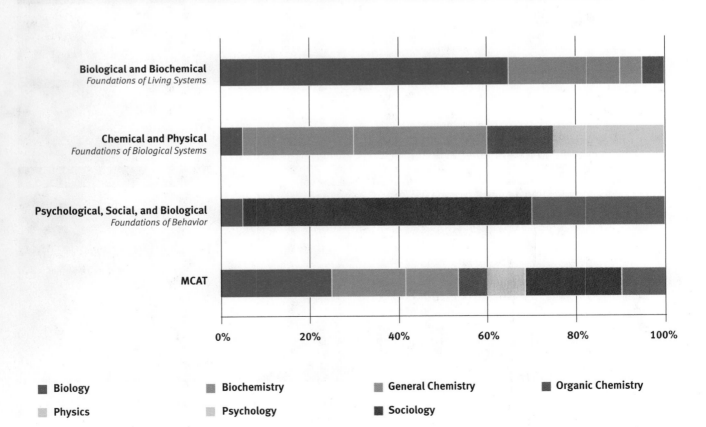

Biological and Biochemical
Foundations of Living Systems

Chemical and Physical
Foundations of Biological Systems

Psychological, Social, and Biological
Foundations of Behavior

MCAT

0%   20%   40%   60%   80%   100%

■ Biology   ■ Biochemistry   ■ General Chemistry   ■ Organic Chemistry

■ Physics   ■ Psychology   ■ Sociology

SCIENTIFIC INQUIRY AND REASONING SKILLS (SIRS)

The AAMC has defined four *Scientific Inquiry and Reasoning Skills* (SIRS) that will be tested in the three science sections of the MCAT:

1. *Knowledge of Scientific Concepts and Principles* (35% of questions)
2. *Scientific Reasoning and Problem-Solving* (45% of questions)
3. *Reasoning About the Design and Execution of Research* (10% of questions)
4. *Data-Based and Statistical Reasoning* (10% of questions)

Let's see how each one breaks down into more specific Test Day behaviors. Note that the bullet points of specific objectives for each of the SIRS are taken directly from the *Official Guide to the MCAT Exam*; the descriptions of what these behaviors mean and sample question stems, however, are written by Kaplan.

Skill 1: *Knowledge of Scientific Concepts and Principles*

This is probably the least surprising of the four SIRS; the testing of science knowledge is, after all, one of the signature qualities of the MCAT. Skill 1 questions will require you to do the following:

- Recognize correct scientific principles
- Identify the relationships among closely related concepts
- Identify the relationships between different representations of concepts (verbal, symbolic, graphic)
- Identify examples of observations that illustrate scientific principles
- Use mathematical equations to solve problems

At Kaplan, we simply call these Science Knowledge or Skill 1 questions. Another way to think of Skill 1 questions is as "one-step" problems. The single step is either to realize which scientific concept the question stem is suggesting or to take the concept stated in the question stem and identify which answer choice is an accurate application of it. Skill 1 questions are particularly prominent among discrete questions (those not associated with a passage). These questions are an opportunity to gain quick points on Test Day—if you know the science concept attached to the question, then that's it! On Test Day, 35% of the questions in each science section will be Skill 1 questions.

Here are some sample Skill 1 question stems:

- How would a proponent of the James–Lange theory of emotion interpret the findings of the study cited the passage?
- Which of the following most accurately describes the function of FSH in the human female menstrual cycle?
- If the products of Reaction 1 and Reaction 2 were combined in solution, the resulting reaction would form:
- Ionic bonds are maintained by which of the following forces?

Skill 2: *Scientific Reasoning and Problem-Solving*

The MCAT science sections do, of course, move beyond testing straightforward science knowledge; Skill 2 questions are the most common way in which it does so. At Kaplan, we also call these Critical Thinking questions. Skill 2 questions will require you to do the following:

- Reason about scientific principles, theories, and models
- Analyze and evaluate scientific explanations and predictions
- Evaluate arguments about causes and consequences
- Bring together theory, observations, and evidence to draw conclusions
- Recognize scientific findings that challenge or invalidate a scientific theory or model
- Determine and use scientific formulas to solve problems

Just as Skill 1 questions can be thought of as "one-step" problems, many Skill 2 questions are "two-step" problems, and more difficult Skill 2 questions may require three or more steps. These questions can require a wide spectrum of reasoning skills, including integration of multiple facts from a passage, combination of multiple science content areas, and prediction of an experiment's results. Skill 2 questions also tend to ask about science content without actually mentioning it by name. For example, a question might describe the results of one experiment and ask you to predict the results of a second experiment without actually telling you what underlying scientific principles are at work—part of the question's difficulty will be figuring out which principles to apply in order to get the correct answer. On Test Day, 45% of the questions in each science section will be Skill 2 questions.

Here are some sample Skill 2 question stems:

- Which of the following experimental conditions would most likely yield results similar to those in Figure 2?
- All of the following conclusions are supported by the information in the passage EXCEPT:
- The most likely cause of the anomalous results found by the experimenter is:
- An impact to a man's chest quickly reduces the volume of one of his lungs to 70% of its initial value while not allowing any air to escape from the man's mouth. By what percentage is the force of outward air pressure increased on a 2 cm^2 portion of the inner surface of the compressed lung?

Skill 3: *Reasoning About the Design and Execution of Research*

The MCAT is interested in your ability to critically appraise and analyze research, as this is an important day-to-day task of a physician. We call these questions Skill 3 or Experimental and Research Design questions for short. Skill 3 questions will require you to do the following:

- Identify the role of theory, past findings, and observations in scientific questioning
- Identify testable research questions and hypotheses
- Distinguish between samples and populations and distinguish results that support generalizations about populations
- Identify independent and dependent variables
- Reason about the features of research studies that suggest associations between variables or causal relationships between them (such as temporality and random assignment)
- Identify conclusions that are supported by research results
- Determine the implications of results for real-world situations
- Reason about ethical issues in scientific research

Over the years, the AAMC has received input from medical schools to require more practical research skills of MCAT test-takers, and Skill 3 questions are the response to these demands. This skill is unique in that the outside knowledge you need to answer Skill 3 questions is not taught in any one undergraduate course; instead, the research design principles needed to answer these questions are learned gradually throughout your science classes and especially through any laboratory work you have completed. It should be noted that Skill 3 comprises 10% of the questions in each science section on Test Day.

Here are some sample Skill 3 question stems:

- What is the dependent variable in the study described in the passage?
- The major flaw in the method used to measure disease susceptibility in Experiment 1 is:
- Which of the following procedures is most important for the experimenters to follow in order for their study to maintain a proper, randomized sample of research subjects?
- A researcher would like to test the hypothesis that individuals who move to an urban area during adulthood are more likely to own a car than are those who have lived in an urban area since birth. Which of the following studies would best test this hypothesis?

Skill 4: *Data-Based and Statistical Reasoning*

Lastly, the science sections of the MCAT test your ability to analyze the visual and numerical results of experiments and studies. We call these Data and Statistical Analysis questions. Skill 4 questions will require you to do the following:

- Use, analyze, and interpret data in figures, graphs, and tables
- Evaluate whether representations make sense for particular scientific observations and data
- Use measures of central tendency (mean, median, and mode) and measures of dispersion (range, interquartile range, and standard deviation) to describe data
- Reason about random and systematic error
- Reason about statistical significance and uncertainty (interpreting statistical significance levels and interpreting a confidence interval)
- Use data to explain relationships between variables or make predictions
- Use data to answer research questions and draw conclusions

Skill 4 is included in the MCAT because physicians and researchers spend much of their time examining the results of their own studies and the studies of others, and it's very important for them to make legitimate conclusions and sound judgments based on that data. The MCAT tests Skill 4 on all three science sections with graphical representations of data (charts and bar graphs) as well as numerical ones (tables, lists, and results summarized in sentence or paragraph form). On Test Day, 10% of the questions in each science section will be Skill 4 questions.

Here are some sample Skill 4 question stems:

- According to the information in the passage, there is an inverse correlation between:
- What conclusion is best supported by the findings displayed in Figure 2?
- A medical test for a rare type of heavy metal poisoning returns a positive result for 98% of affected individuals and 13% of unaffected individuals. Which of the following types of error is most prevalent in this test?
- If a fourth trial of Experiment 1 was run and yielded a result of 54% compliance, which of the following would be true?

SIRS Summary

Discussing the SIRS tested on the MCAT is a daunting prospect given that the very nature of the skills tends to make the conversation rather abstract. Nevertheless, with enough practice, you'll be able to identify each of the four skills quickly, and you'll also be able to apply the proper strategies to solve those problems on Test Day. If you need a quick reference to remind you of the four SIRS, these guidelines may help:

Skill 1 (Science Knowledge) questions ask:

- Do you remember this science content?

Skill 2 (Critical Thinking) questions ask:

- Do you remember this science content? And if you do, could you please apply it to this novel situation?
- Could you answer this question that cleverly combines multiple content areas at the same time?

Skill 3 (Experimental and Research Design) questions ask:

- Let's forget about the science content for a while. Could you give some insight into the experimental or research methods involved in this situation?

Skill 4 (Data and Statistical Analysis) questions ask:

- Let's forget about the science content for a while. Could you accurately read some graphs and tables for a moment? Could you make some conclusions or extrapolations based on the information presented?

CRITICAL ANALYSIS AND REASONING SKILLS (CARS)

The *Critical Analysis and Reasoning Skills* (CARS) section of the MCAT tests three discrete families of textual reasoning skills; each of these families requires a higher level of reasoning than the last. Those three skills are as follows:

1. *Foundations of Comprehension* (30% of questions)
2. *Reasoning Within the Text* (30% of questions)
3. *Reasoning Beyond the Text* (40% of questions)

These three skills are tested through nine humanities- and social sciences–themed passages, with approximately 5 to 7 questions per passage. Let's take a more in-depth look into these three skills. Again, the bullet points of specific objectives for each of the CARS are taken directly from the *Official Guide to the MCAT Exam*; the descriptions of what these behaviors mean and sample question stems, however, are written by Kaplan.

Foundations of Comprehension

Questions in this skill will ask for basic facts and simple inferences about the passage; the questions themselves will be similar to those seen on reading comprehension sections of other standardized exams like the SAT® and ACT®. *Foundations of Comprehension* questions will require you to do the following:

- Understand the basic components of the text
- Infer meaning from rhetorical devices, word choice, and text structure

This admittedly covers a wide range of potential question types including Main Idea, Detail, Function, and Definition-in-Context questions, but finding the correct answer to all *Foundations of Comprehension* questions will follow from a basic understanding of the passage and the point of view of its author (and occasionally that of other voices in the passage).

Here are some sample *Foundations of Comprehension* question stems:

- **Main Idea**—The author's primary purpose in this passage is:
- **Detail**—Based on the information in the second paragraph, which of the following is the most accurate summary of the opinion held by Schubert's critics?
- **(Scattered) Detail**—According to the passage, which of the following is FALSE about literary reviews in the 1920s?
- **Function**—The author's discussion of the effect of socioeconomic status on social mobility primarily serves which of the following functions?
- **Definition-in-Context**—The word "obscure" (paragraph 3), when used in reference to the historian's actions, most nearly means:

Reasoning Within the Text

While *Foundations of Comprehension* questions will usually depend on interpreting a single piece of information in the passage or understanding the passage as a whole, *Reasoning Within the Text* questions will typically require you to infer unstated parts of arguments or bring together two disparate pieces of the passage. *Reasoning Within the Text* questions will require you to:

- Integrate different components of the text to increase comprehension

In other words, questions in this skill often ask either *How do these two details relate to one another?* or *What else must be true that the author didn't say?* The CARS section will also ask you to judge certain parts of the passage or even judge the author. These questions, which fall under the *Reasoning Within the Text* skill, can ask you to identify authorial bias, evaluate the credibility of cited sources, determine the logical soundness of an argument, or search for relevant evidence in the passage to support a given conclusion. In all, this category includes Inference and Strengthen–Weaken (Within the Passage) questions, as well as a smattering of related—but rare—question types.

Here are some sample *Reasoning Within the Text* question stems:

- **Inference (Implication)**—Which of the following phrases, as used in the passage, is most suggestive that the author has a personal bias toward narrative records of history?
- **Inference (Assumption)**—In putting together her argument in the passage, the author most likely assumes:
- **Strengthen–Weaken (Within the Passage)**—Which of the following facts is used in the passage as the most prominent piece of evidence in favor of the author's conclusions?
- **Strengthen–Weaken (Within the Passage)**—Based on the role it plays in the author's argument, *The Possessed* can be considered:

Reasoning Beyond the Text

The distinguishing factor of *Reasoning Beyond the Text* questions is in the title of the skill: the word *Beyond*. Questions that test this skill, which make up a larger share of the CARS section than questions from either of the other two skills, will always introduce a completely new situation that was not present in the passage itself; these questions will ask you to determine how one influences the other. *Reasoning Beyond the Text* questions will require you to:

- Apply or extrapolate ideas from the passage to new contexts
- Assess the impact of introducing new factors, information, or conditions to ideas from the passage

The *Reasoning Beyond the Text* skill is further divided into Apply and Strengthen–Weaken (Beyond the Passage) questions, and a few other rarely appearing question types.

Here are some sample *Reasoning Beyond the Text* question stems:

- **Apply**—If a document were located that demonstrated Berlioz intended to include a chorus of at least 700 in his *Grande Messe des Mortes*, how would the author likely respond?
- **Apply**—Which of the following is the best example of a "virtuous rebellion," as it is defined in the passage?
- **Strengthen–Weaken (Beyond the Text)**—Suppose Jane Austen had written in a letter to her sister, "My strongest characters were those forced by circumstance to confront basic questions about the society in which they lived." What relevance would this have to the passage?
- **Strengthen–Weaken (Beyond the Text)**—Which of the following sentences, if added to the end of the passage, would most WEAKEN the author's conclusions in the last paragraph?

CARS Summary

Through the *Foundations of Comprehension* skill, the CARS section tests many of the reading skills you have been building on since grade school, albeit in the context of very challenging doctorate-level passages. But through the two other skills (*Reasoning Within the Text* and *Reasoning Beyond the Text*), the MCAT demands that you understand the deep structure of passages and the arguments within them at a very advanced level. And, of course, all of this is tested under very tight timing restrictions: only 102 seconds per question—and that doesn't even include the time spent reading the passages.

Here's a quick reference guide to the three CARS skills:

Foundations of Comprehension questions ask:

- Did you understand the passage and its main ideas?
- What does the passage have to say about this particular detail?

Reasoning Within the Text questions ask:

- What must be true that the author did not say?
- What's the logical relationship between these two ideas from the passage?
- How well argued is the author's thesis?

Reasoning Beyond the Text questions ask:

- How does this principle from the passage apply to this new situation?
- How does this new piece of information influence the arguments in the passage?

Scoring

Each of the four sections of the MCAT is scored between 118 and 132, with the median at 125. This means the total score ranges from 472 to 528, with the median at 500. Why such peculiar numbers? The AAMC stresses that this scale emphasizes the importance of the central portion of the score distribution, where most students score (around 125 per section, or 500 total), rather than putting undue focus on the high end of the scale.

Note that there is no wrong answer penalty on the MCAT, so you should select an answer for every question—even if it is only a guess.

The AAMC has released a hypothetical correlation between scaled score and percentile, as shown here. It should be noted that the percentile scale is adjusted and renormalized over time and thus should not be overinterpreted.

Total Score	Percentile	Total Score	Percentile
528	>99	499	49
527	>99	498	45
526	>99	497	42
525	>99	496	39
524	>99	495	35
523	>99	494	32
522	99	493	29
521	99	492	26
520	98	491	23
519	98	490	20
518	97	489	18
517	96	488	16
516	95	487	13
515	94	486	12
514	92	485	10
513	90	484	8
512	88	483	7
511	86	482	5

Total Score	Percentile	Total Score	Percentile
510	84	481	4
509	82	480	3
508	79	479	2
507	76	478	2
506	73	477	1
505	70	476	1
504	67	475	<1
503	63	474	<1
502	60	473	<1
501	56	472	<1
500	53		

Source: AAMC. 2016. *Summary of MCAT Total and Section Scores*. Accessed December 2016. **https://students-residents.aamc.org/advisors/article/percentile-ranks-for-the-mcat-exam/**.

Further information on score reporting is included at the end of the next section (see *After Your Test*).

MCAT POLICIES AND PROCEDURES

We strongly encourage you to download the latest copy of *MCAT® Essentials*, available on the AAMC's website, to ensure that you have the latest information about registration and Test Day policies and procedures; this document is updated annually. A brief summary of some of the most important rules is provided here.

MCAT Registration

The only way to register for the MCAT is online. You can access AAMC's registration system at: **www.aamc.org/mcat**.

You will be able to access the site approximately six months before Test Day. The AAMC designates three registration "Zones"—Gold, Silver, and Bronze. Registering during the Gold Zone (from the opening of registration until approximately one month before Test Day) provides the most flexibility and lowest test fees. The Silver Zone runs until approximately two to three weeks before Test Day and has less flexibility and higher fees; the Bronze Zone runs until approximately one to two weeks before Test Day and has the least flexibility and highest fees.

Fees and the Fee Assistance Program (FAP)

Payment for test registration must be made by MasterCard or VISA. As described earlier, the fees for registering for the MCAT—as well as rescheduling the exam or changing your testing center—increase as one approaches Test Day. In addition, it is not uncommon for test centers to fill up well in advance of the registration deadline. For these reasons, we recommend identifying your preferred Test Day as soon as possible and registering. There are ancillary benefits to having a set Test Day, as well: when you know the date you're working toward, you'll study harder and are less likely to keep pushing back the exam. The AAMC offers a Fee Assistance Program (FAP) for students with financial hardship to help reduce the cost of taking the MCAT, as well as for the American Medical College Application Service (AMCAS®) application. Further information on the FAP can be found at: **www.aamc.org/students/applying/fap**.

Testing Security

On Test Day, you will be required to present a government-issued ID. When registering, take care to spell your name precisely the same as it appears on this ID; failure to provide this ID at the test center or differences in spelling between your registration and ID will be considered a "no-show," and you will not receive a refund for the exam.

You will also be required to provide an electronic thumbprint and electronic signature verification to take the exam. Some testing centers may use a metal detection wand to ensure that no prohibited items are brought into the testing room. Prohibited items include all electronic devices, including watches and timers, calculators, cell phones, and any and all forms of recording equipment; food, drinks (including water), and cigarettes or other smoking paraphernalia; hats and scarves (except for religious purposes); and books, notes, or other study materials. If you require a medical device, such as an insulin pump or pacemaker, you must apply for accommodated testing. During breaks, you are allowed to access food and drink, but not electronic devices, including cell phones.

Testing centers are under video surveillance and the AAMC does not take potential violations of testing security lightly. The bottom line: *know the rules and don't break them.*

Accommodations

Students with disabilities or medical conditions can apply for accommodated testing. Documentation of the disability or condition is required, and requests may take two months—or more—to be approved. For this reason, it is recommended that you begin the process of applying for accommodated testing as early as possible. More information on applying for accommodated testing can be found at: **www.aamc.org/students/applying/mcat/accommodations**.

After Your Test

When your MCAT is all over, no matter how you feel you did, be good to yourself when you leave the test center. Celebrate! Take a nap. Watch a movie. Ride your bike. Plan a trip. Call up all of your neglected friends or stalk them on Facebook. Totally consume a cheesesteak and drink dirty martinis at night (assuming you're over 21). Whatever you do, make sure that it has absolutely nothing to do with thinking too hard—you deserve some rest and relaxation.

Perhaps most importantly, do not discuss specific details about the test with anyone. For one, it is important to let go of the stress of Test Day, and reliving your exam only inhibits you from being able to do so. But more significantly, the Examinee Agreement you sign at the beginning of your exam specifically prohibits you from discussing or disclosing exam content. The AAMC is known to seek out individuals who violate this agreement and retains the right to prosecute these individuals at their discretion. This means that you should not, under any circumstances, discuss the exam in person or over the phone with other individuals—including us at Kaplan—or post information or questions about exam content to Facebook, Student Doctor Network, or other online social media. You are permitted to comment on your "general exam experience," including how you felt about the exam overall or an individual section, but this is a fine line. In summary: *if you're not certain whether you can discuss an aspect of the test or not, just don't do it!* Do not let a silly Facebook post stop you from becoming the doctor you deserve to be.

Scores are released approximately one month after Test Day. The release is staggered during the afternoon and evening, starting at 5 p.m. Eastern. This means that not all examinees receive their scores at exactly the same time. Your score report will include a scaled score for each section between 118 and 132, as well as your total combined score between 472 and 528. These scores are given as confidence intervals. For each section, the confidence interval is approximately the given score ± 1; for the total score, it is approximately the given score ± 2. You will also be given the corresponding percentile rank for each of these section scores and the total score.

AAMC CONTACT INFORMATION

For further questions, contact the MCAT team at the Association of American Medical Colleges:

MCAT Resource Center
Association of American Medical Colleges
www.aamc.org/mcat
(202) 828-0690
mcat@aamc.org

How This Book Was Created

This book began as a book called the *High-Yield Problem-Solving Guide* (HYPSG). The *High-Yield Problem-Solving Guide* project began in November 2012 shortly after the release of the *Preview Guide for the MCAT 2015 Exam,* 2nd edition. Through thorough analysis by our staff psychometricians, we were able to analyze the relative yield of the different topics on the MCAT and began constructing tables of contents for the books of the *Kaplan MCAT Review* series to which the HYPSG is related.

Writing of the books began in April 2013. A dedicated staff of 30 writers, 7 editors, and 32 proofreaders worked over 5000 combined hours to produce these books. The format of the books was heavily influenced by weekly meetings with Kaplan's learning science team.

In 2016, several of the problems from the original *High-Yield Problem Solving Guide* were reworked and incorporated into the core Kaplan MCAT class. These problems are the ones marked "Solutions" throughout this book. The remaining problems were retained and are now marked as "Extra Practice." With that change, this book became a companion to the core Kaplan MCAT class, and it was retitled *High-Yield Science: Solutions and Extra Practice* to reflect its new function in the course. The text was further refined through revision in 2017.

This book was submitted for publication in June 2017. For any updates after this date, please visit www.kaplanmcat.com.

Each question in this book has been vetted through at least ten rounds of review. To that end, the information presented in these books is true and accurate to the best of our knowledge. Still, your feedback helps us improve our prep materials. Please notify us of any inaccuracies or errors in the books by sending an email to **KaplanMCATfeedback@kaplan.com**.

Behavioral Sciences

Key Concepts

Biology Chapter 1

Sympathetic nervous system

Parasympathetic nervous system

Somatic function

Autonomic function

Reflex arc

Organization of the Human ⑤ Nervous System

Henry is a newborn infant who has been taken to his physician for a checkup. Up until now, Henry has behaved normally. His doctor assesses several things to ensure that Henry is healthy.

- The doctor asks Henry's parent whether Henry is either sleeping or becoming more docile after nursing.
- The doctor tests Henry's vision and range of motion by waving a brightly colored object in front of Henry and watching his response, which should typically involve visual tracking of the object and an attempt to move extremities towards or away from the object.
- The doctor tests Henry's Galant reflex. In this test the physician taps on the infant's spine, and the infant's hips should involuntarily and immediately twitch toward the location of the tap.

Henry behaves as expected in the first two tasks, but fails to exhibit the Galant reflex. What should the physician focus on in further neurological assessment?

① **How does the first test give information about peripheral nervous system function?**

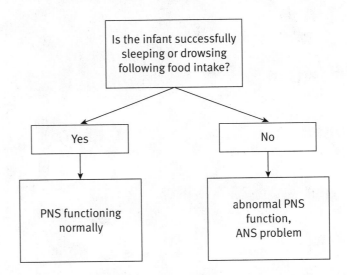

Activation of the peripheral nervous systems falls into two major categories: somatic and autonomic function. Somatic function is limited to voluntary muscle motion. Autonomic function includes both sympathetic and parasympathetic responses. If the parasympathetic branch of the nervous system is functioning properly, then food intake should be paired with digestion and decreased arousal. This response is referred to as the "rest and digest" response. If Henry is successfully sleeping or drowsing after eating, that information indicates that the parasympathetic branch of his peripheral nervous system is responding appropriately to incoming stimuli. If Henry was instead becoming more aroused or agitated after eating, this information would indicate potential disruption of autonomic nervous system function and a potential issue in Henry's peripheral nervous system.

 How does the bright object test assess both major divisions of the nervous system?

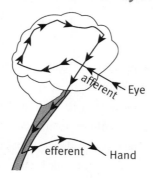

Central Nervous System = Brain and Spinal Cord
Peripheral Nervous System = Sensory Receptors and Motor Neurons

The bright object test begins with exposing Henry to a bright object he observes. The processing of environmental information via sensory neurons in the eye is a function of the peripheral nervous system. In order to process visual data, that sensory information must pass through several regions of the brain. Afferent innervation must travel via the optic nerve through the lateral geniculate nucleus to the visual cortex in the occipital lobe for production of a usable image. After the visual data is processed, it must be interpreted by the frontal cortex, and any response by the infant will be generated via the motor cortex. Once a response has been determined, that feedback must be passed downward through the spinal cord. Ultimately, any movement toward or away from the object requires innervation of the muscles, which involves signals leaving the spinal cord and acting on the muscles of the extremities. The bright object test can thus allow for assessment of several aspects of the nervous system at once.

Takeaways

The nervous system is organized by evolutionary order of development, so the complexity of function generally increases as you move upward through the central nervous system. Notice that the peripheral nervous system is responsible for carrying out the instructions that result from the more advanced signaling systems within the brain.

Things to Watch Out For

It is easy to confuse the terminology of the nervous system. While autonomic function is part of the peripheral nervous system, the peripheral nervous system also includes somatic function. Though the brain is a major component of the central nervous system, not all CNS function occurs within the brain, as the spinal cord is also included in this category.

3 What does the Galant reflex test provide information about?

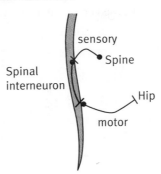

The Galant reflex is a reflex arc, involving afferent input to a sensory neuron located along the spine, followed by a spinal cord reflex interneuron, and then efferent output via a motor neuron on the hip. This test thus provides information about the function of spinal column interneuron function, spinal sensory neurons, and hip motor neurons. The Galant reflex test does not provide any information about the function of the brain or the autonomic branch of the peripheral nervous system.

4 Which potentially systemic issue should the physician look for first?

Henry is an infant who has already successfully demonstrated proper function of motor neurons and sensory neurons via the bright object test. Additionally, it is unlikely that any systemic issues with sensory or motor function would go unnoticed in an infant prior to testing. The only other system that the Galant reflex test informs the physician about is spinal reflex interneuron function. The doctor has not performed any other tests that assess reflex interneurons, so these interneurons would be the best target for further testing to eliminate systemic issues.

Related Questions

1. What are the parts of the hindbrain and the meaningful functions these areas serve?

2. What are the three parts of the hypothalamus and what functions does each part serve?

3. What parts of the brain make up the cerebrum? How is the cerebrum associated with Parkinson's disease and schizophrenia?

High-Yield Problem-Solving Guide questions continue on the next page. ▶ ▶ ▶

E Vision

Multiple sclerosis (MS) is a common demyelinating disorder that affects the white matter of the central nervous system. The visual pathways are commonly affected; optic neuritis, which is inflammation of the optic nerve, is one of the most common presentations. Less commonly, MS may cause lesions of an entire section of the visual pathway, leading to visual field defects. What would be the visual field defect in an individual who lost function of the right optic nerve? Of the fibers crossing in the optic chiasm? Of the right optic tract? Of the right occipital lobe?

1 What fibers are carried in the right optic nerve?

The right optic nerve collects the axons of neurons leaving the retina of the right eye. The nasal visual field (toward the nose) projects to the temporal retina (toward the side of the head). Similarly, the temporal visual field projects to the nasal retina. Together, all of these fibers form the right optic nerve.

If the entire right optic nerve were lesioned in MS, then all of information from the right eye would be lost. In medicine, this may be represented as shown:

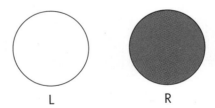

When reading visual fields, imagine that you are looking out at the world through the two circles; in this example, nothing is visible through the right eye.

2 What fibers cross in the optic chiasm?

In the optic chiasm, the nasal fibers cross while the temporal fibers pass directly through to the ipsilateral (on the same side) optic tract. Remember that the nasal fibers carry information from the temporal visual field. Therefore, a lesion of the fibers crossing in the optic chiasm would knock out nasal fibers from both eyes and cause a loss of the temporal visual field from both eyes. This is called bitemporal hemianopsia, and can be represented as shown:

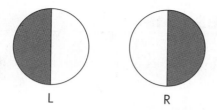

3 What fibers are carried in the right optic tract?

Because of the crossing in the optic chiasm, the right optic tract carries the nasal fibers from the left eye and the temporal fibers from the right eye. These correspond to the temporal visual field from the left eye and the nasal visual field from the right eye, respectively—which is the left visual field of each eye. A complete lesion of the right optic tract would therefore cause a loss of the left visual field from both eyes. This is called homonymous hemianopsia and can be represented as shown:

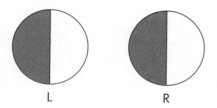

4 What fibers are carried in the right occipital lobe?

The optic tracts project directly to their respective occipital lobes. Therefore, damage to the right occipital lobe will cause loss of the left visual field. Interestingly, occipital lobe lesions (which occur commonly during a stroke) are often marked by macular sparing, or survival of the central visual field in each eye, and can be represented as shown:

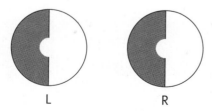

This is due to differences in circulation to the different parts of the occipital lobe, which is outside the scope of the MCAT.

Takeaways

An optic nerve carries all the information from the associated eye. The optic chiasm contains crossing nasal fibers (the temporal visual field) from each eye. An optic tract carries all the information from the opposite visual field.

Things to Watch Out For

It is very easy to misread a question about the visual system—does the answer refer to a part of the visual pathway or to a part of the visual field? Wrong answers often simply mix up *left* and *right* or *nasal* and *temporal*.

Related Questions

1. What are the two main types of photoreceptors in the retina? How are these photoreceptor types distributed?

2. What is feature detection theory?

3. What is parallel processing?

High-Yield Problem-Solving Guide questions continue on the next page. ▶ ▶ ▶

ⓈAssociative Learning

A pigeon is placed in a cage with two buttons: one blue and one green. Pressing the blue button triggers the immediate release of a food pellet. However, pressing the blue button also immediately triggers an electric shock. Furthermore, each time a food pellet is released, the mechanism used to release food pellets is disabled for a period of time varying randomly between 5 and 55 seconds. The blue button's electric shock mechanism remains active during this period. And once the latency period has passed, the next press of the blue button will immediately trigger the release of another food pellet. Finally, pressing the green button will disable the electric shock mechanism until the next food pellet is released, at which point the electric shock mechanism will reactivate. What pattern of button presses will a trained pigeon use to successfully receive a food pellet without being shocked?

1 **What makes the blue button both positive reinforcement and ALSO positive punishment?**

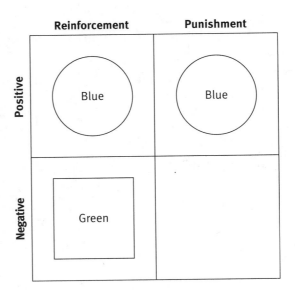

Pressing the blue button triggers two consequences. One of these consequences is the electrical shock, which is used as positive punishment because the shock adds an unpleasant stimulus that is meant to deter behavior. Remember that "positive" refers to the addition of a stimulus (the shock) and "punishment" refers to the intended outcome (to decrease behavior). The other consequence of pressing the blue button—the food pellet—is positive reinforcement because the food pellet is a desirable result for the bird. Remember that "reinforcement" refers to the intended outcome (to increase behavior).

2 What makes the green button negative reinforcement?

Pressing the green button removes an unpleasant stimulus, making the green button a form of negative reinforcement. "Negative" refers to the removal of a stimulus (the shock). In addition, because the green button prevents the shock from occurring, pressing it is also an example of avoidance learning.

3 What about the blue button's reward makes it a variable interval schedule?

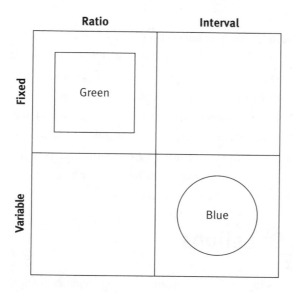

The blue button's release of a food pellet occurs after a latency period of time varying randomly between 5 and 55 seconds, meaning it is dependent on some amount of time passing. Such a latency period is the key trademark of an interval schedule. Given that length of the latency period varies, this reinforcement schedule is a variable interval schedule.

4 What about the green button's removal of punishment makes it a fixed ratio schedule?

"Ratio" is the term used when reinforcement is dependent on the number of button presses, not on the passage of time. When the number of presses required for reinforcement stays the same, as is this case with a reinforcement being delivered consistently after just a single button press, the ratio is fixed.

5 **After the pigeon has been conditioned by the buttons, how will it use them over time?**

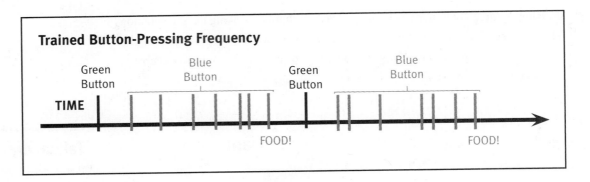

Trained Button-Pressing Frequency

Green Button   Blue Button   Green Button   Blue Button

TIME

FOOD!   FOOD!

As the food reward for the blue button is on a variable interval schedule, you should expect the frequency of button presses to be relatively low but constant. With enough training, the pigeon will learn to estimate the average amount of time required to earn a reward, and will learn to press the blue button more frequently if that average time is exceeded. A button on a fixed-ratio schedule would normally be pressed more frequently than on a variable-interval schedule, but in the scenario described, the green button only affects the consequences of the blue button. We would therefore expect the pigeon to press the green button only as needed: once per interval immediately after the food reward from the previous interval.

Related Questions

1. Emily wishes to stop her cat from knocking items off of a shelf, but doesn't believe in punishment. How can she use operant conditioning to decrease this behavior without punishing the cat?

2. A fetish is a nonsexual behavior or object that becomes sexualized and is often able to provide sexual satisfaction on its own. How might the principles of associative learning help to explain fetish development?

3. Ted's favorite restaurant plays the same music each day during lunch. He realizes that whenever he hears one of the songs, he feels hungry. Describe the classical conditioning stimuli and responses in this scenario. What will happen if he stops going to the restaurant?

High-Yield Problem-Solving Guide questions continue on the next page. ▶ ▶ ▶

E Language

A 73-year-old male patient has recently had a stroke. While he appears to be recovering and is able to follow verbal instructions, he has lost the ability to speak. What language disorder has this patient likely developed? What part of the patient's brain has likely been damaged to cause this disorder?

1 What is the scope of possible language disorders in this patient?

Because the patient has had a stroke, he most likely has developed aphasia— a disturbance in the ability to comprehend or formulate language resulting from dysfunction in a specific brain region. Three major forms of aphasia are Broca's aphasia, Wernicke's aphasia, and conduction aphasia.

2 What are the symptoms of the different types of aphasia?

When one assesses aphasia, the three most important characteristics to look for are speech comprehension, speech production, and the ability to repeat verbal information.

Speech comprehension is primarily the responsibility of the temporal lobe; more specifically, Wernicke's area on the superior temporal gyrus is involved in interpreting the content of speech. Other regions of the temporal lobe are important for interpreting other auditory stimuli, such as incident sound and music.

Speech production is primarily the responsibility of the frontal lobe; more specifically, Broca's area on the inferior frontal gyrus is involved in the coordination of the muscles of the larynx, pharynx, palate, tongue, and lips that allow for the production of speech.

The ability to repeat verbal information is the combined responsibility of Wernicke's area, Broca's area, and the arcuate fasciculus—a band of white matter that connects the two areas. While all three areas must be intact to permit repetition of speech, an isolated loss of the ability to repeat verbal information (conduction aphasia) may result from isolated damage to the arcuate fasciculus.

What symptoms does this patient exhibit?

This patient has intact speech comprehension, as evidenced by his ability to follow verbal commands, but has difficulty with speech production. We do not have enough information to assess his ability to repeat verbal information; however, even if a clinician tested for repetition ability, the results would be obscured by the difficulty with speech production.

 ### What form of aphasia does the patient most likely have?

The inability to produce speech, combined with intact speech comprehension, is characteristic of Broca's aphasia. Broca's aphasia results from damage to Broca's area; this is a very common lesion in strokes because it falls in the vascular distribution of the middle cerebral artery—the most frequently affected vessel in a stroke.

It is possible that this patient also suffered damage to the nearby arcuate fasciculus, but we do not have enough information to draw this conclusion. We have strong evidence for Broca's aphasia, which would mask the symptoms of conduction aphasia.

Related Questions

1.   What is the typical timeline for language development in children?
2.   What is the nativist theory of language acquisition?
3.   What is the learning theory of language acquisition?

Takeaways

Broca's aphasia includes intact speech comprehension but reduced or absent speech production; Wernicke's aphasia includes intact speech production but reduced or absent speech comprehension; conduction aphasia includes intact speech production and comprehension but the loss of the ability to repeat verbal information.

Things to Watch Out For

Many cases of aphasia can incorporate attributes of the different diagnoses described in this question; major damage to both the frontal and temporal lobes could lead to loss of both speech production and speech comprehension, for example.

⑤ Theories of Emotion

One major early attempt to explain the relationship between physiological events and emotion was made in the 19th century by William James and Carl Lange: the James–Lange theory of emotion. This theory held that a precipitating event would lead to physiological arousal, which would be followed by neural interpretation and generation of an emotional response. This (now largely defunct) theory has been strongly refuted by later research, especially that of Walter Cannon and Philip Bard. These scientists ran a series of studies in which sympathetic nervous system structures were completely removed from animal models, and stressful situations were then used to attempt to induce emotional response. Despite the elimination of this source of physiological input, the animal models in these studies still demonstrated emotional responses to precipitating events. In later experiments, subjects' viscera (when present, but detached from direct brain connection) still experienced physiological arousal in the stressful situations created by the experimenters. Based on this research and its findings, how would the Cannon–Bard theory be expected to explain the generation of emotion?

① How does the Cannon–Bard removal of sympathetic feedback demonstrate a flaw within the James–Lange theory?

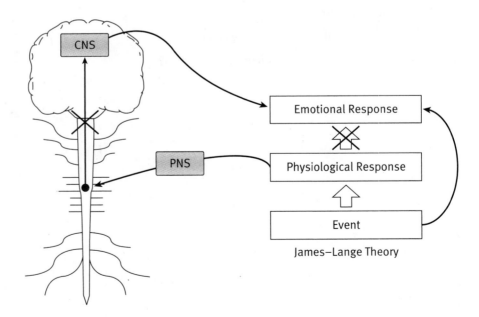

James–Lange Theory

The James–Lange theory holds that a precipitating event directly leads to a physiological response. In this theory, the physiological response, typically activation of the autonomic nervous system, signals to the central nervous system and emotion-generating brain regions. Accordingly, James–Lange holds that the physiological response must come before the emotional response, as the physiological response is required for the emotional response to occur. Cannon and Bard, in their subsequent research, removed autonomic input to the central nervous system, but Cannon and Bard still saw emotion generated in test subjects. The research of Cannon and Bard thus demonstrates that the physiological response is not necessary for the generation of the emotional response.

2 **In the Cannon–Bard early experiments, animals experienced emotion without involving any physiological response. Why wouldn't they remove physiology from the theory of emotion?**

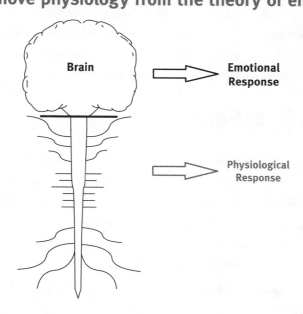

The Cannon–Bard experiment from the question stem removed physiological input to the CNS, but still saw emotion generated by the CNS. The research successfully demonstrated that physiological input was not required as a response prior to emotional output. However, this research did not test whether or not physiological arousal would occur in response to a stimulating event. In fact, the researchers remove the viscera entirely, preventing any testing of the physiological response. Given that physiological response is a known response to a precipitating event, as seen in the later Cannon–Bard research, physiological response should still follow the precipitating event in the Cannon–Bard theory. Further, it is possible that physiological response does modulate emotional response, but does not need to precede generation of emotion—yet another reason why physiology cannot be removed from the Cannon–Bard model.

Takeaways

The Cannon–Bard theory of emotion states that an emotion and a physiological response result from a stimulus. Neither of these responses relies on the other.

Things to Watch Out For

Make sure you understand the subtle differences between the James–Lange, Cannon–Bard, and Schachter–Singer theories of emotion—all three play a part in our current understanding of how emotions arise. Also remember that the spinal cord is not the only pathway for feedback from the visceral organs—the vagus nerve, which feeds directly into the medulla oblongata, performs this function as well.

3 **Given the flaw found in the James–Lange theory, what should the Cannon–Bard theory look like?**

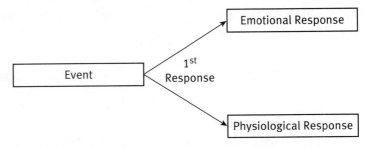

The research of Cannon and Bard successfully demonstrated that the order of events in the James–Lange theory was unnecessary and inaccurate in their animal model. However, Cannon and Bard still saw a physiological response following the precipitating event, and this response occurred independently of the emotional response. The Cannon–Bard theory shows both physiology and emotion following the precipitating event, though there is no set order of events between the two.

Related Questions

1. A car cuts you off as you're driving down a highway. As a result, you feel angry. You also notice that your heart rate is elevated, your mouth feels dry, and your skin feels hot. How does the Cannon–Bard theory of emotion account for the anger you feel and your physiological response?

2. Given the same event as in Related Question 1, what is the sequence of events in the Schachter–Singer theory of emotion?

3. What did Darwin have to say about emotions? How did Ekman further develop Darwin's ideas?

High-Yield Problem-Solving Guide questions continue on the next page. ▶ ▶ ▶

🆂 Stages of Moral Reasoning

Marvin is in his residency program for internal medicine. He encounters a patient whom he identifies as someone who can be helped by Varvipax, a new and expensive drug for her condition. However, the patient's insurance does not cover the drug, and she says she couldn't afford Varvipax at its list price. Rather than prescribe an older, less-effective drug, Marvin gives the patient dozens of the "free samples" of Varvipax he has received from the drug company—enough for a full course of treatment—despite the fact that this is against his office's policy and is against his contract with the drug company. How might Marvin justify this behavior in each of the major stages of Kohlberg's theory of moral development?

1 **If Marvin acted to relieve guilt, what stage of moral reasoning is he in?**

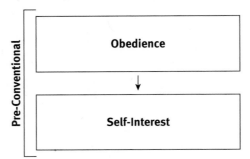

Kohlberg's theory does not focus on the ultimate decision made by an individual (in this case, Marvin), but rather the route by which the individual arrives at his decision. Marvin's reasoning for giving the excess of samples to the patient stems here from a desire to avoid feeling guilt. If Marvin views guilt as a punishment, we would say his reasoning falls into Stage 1: obedience, the phase in which moral reasoning is based on fear of punishment. If instead we had been told that Marvin was gaining some sort of benefit from acting to help the patient, his reasoning would fall under Stage 2: self-interest.

2 **If Marvin acted to be a certain type of person, what stage of moral reasoning is he in?**

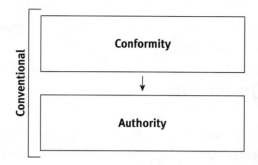

Marvin's reasoning in this situation now stems from his desire to have the reputation as "a doctor who takes care of his patients, no matter what." Marvin's new reasoning here follows conventional reasoning, which involves judging what is morally correct by society's conventions of right and wrong. Marvin is acting here to be thought of in a certain way by his peers. This type of reasoning comes from Stage 3: conformity. Marvin is making his decision in order to conform to the rules he believes society to have concerning his social role. If Marvin's reasoning had instead involved treating the patient in order to obey laws and/or rules, such as the Hippocratic Oath, his reasoning would fall into Stage 4: authority and social order.

3 **If Marvin acted based on the treatment he'd want to receive, what stage of moral reasoning is he in?**

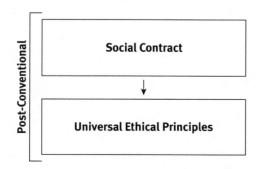

In this final case, Marvin gave the patient the samples "because if the roles were reversed, he would want someone else to give him medicine that he needed but couldn't afford." Marvin's reasoning matches with stage 5: social contract. Marvin is not acting based on current laws, but instead on what he personally believes the moral behavior would be in the situation. In some cases, this means agreement with the law, but in some it means a contradiction with the law, as seen here. As stage 5 reasoning,

Takeaways

Keep in mind that the conclusion of an argument is irrelevant to determining the stage of moral reasoning. Lois and Marvin could have come to opposite conclusions and still have fallen in the same stages of moral reasoning.

Things to Watch Out For

While the differences between phases of moral reasoning are fairly clear-cut, the differences between the numbered stages can be subtle. Make sure to be able to describe the six stages clearly, and carefully examine the content of the arguments made in any moral reasoning question.

Marvin is also aware he's acting for his own moral ideals, not implementing ideals he dictates that everyone should have. If Marvin were in Stage 6: universal ethical principles, he would be acting because he believes his actions represent some universal moral principle. Further, in Stage 6, Marvin's reasoning would include that all should follow this same principle. Kohlberg holds that individuals reasoning at the universal ethical principles level are exceedingly rare.

Related Questions

1. A three-year-old child has begun to explore her surroundings and is learning that she is able to control the world around her. She brushes her own teeth, dresses herself, and has begun to develop personal interests. According to Erikson's stages of psychosocial development, what is the next conflict this child will face?

2. Two college roommates are constantly at odds because one is extremely messy and the other is obsessively tidy. According to Freudian analysis, how can the differences between these roommates be explained in terms of psychosexual fixation?

3. A child of busy parents is often left on his own to play and learn. According to Vygotsky, what effect would the consistent absence of a more knowledgeable other have on the child's locus of control and self-efficacy?

High-Yield Problem-Solving Guide questions continue on the next page. ▶ ▶ ▶

⊟ Bipolar and Depressive Disorders

A 46-year-old woman is brought by her husband to the emergency room after she was found on the roof of her home, yelling incomprehensible statements at the sky. Her husband says that, over the last two weeks, she has been staying up late at night, shopping for expensive clothes that she cannot afford, and sleeping only two to three hours per night. She cannot stop talking and moves rapidly from one topic to another. When asked how she feels, she exclaims, *I'm fantastic!* Last month, she was diagnosed with depression and started on a medication for her symptoms. What diagnosis is she most likely to have?

1 What mood is the patient exhibiting?

The symptoms described here are consistent with mania. Bipolar and depressive disorders were historically classified under the same umbrella of mood disorders in the previous version of the *Diagnostic and Statistical Manual of Mental Disorders* (DSM), the DSM-IV-TR. In the DSM-5, these two types of disorders have been split into separate diagnostic categories. They still share some characteristics, however, so it is useful to think of them together.

Bipolar disorders—at least their manic components—are characterized by elevated, expansive, or irritable mood. In other words, these individuals feel on top of the world or feel irritable. Depressive disorders are characterized by sadness, lack of interest, or apathy. The patient in this example clearly has an elevated mood, as evidenced by her feeling *fantastic!*

2 Do the manic symptoms constitute a manic episode?

According to the DSM-5, a manic episode must include an elevated, expansive, or irritable mood as described earlier and must last for at least one week (although the duration does not matter if the episode is severe enough to warrant hospitalization). The patient must also have at least three of the following symptoms (four if the mood is only irritable):

- Easily **D**istracted
- Decreased need for sleep (**I**nsomnia)
- Increased self-esteem or **G**randiosity
- **F**light of ideas or racing thoughts

- Increased goal-directed activity or motor **A**gitation
- Increased talkativeness or pressured **S**peech
- Involvement in high-risk behavior (**T**houghtlessness)

These symptoms must cause significant impairment, require hospitalization, or be accompanied by psychosis. The mnemonic **DIGFAST** is a simple way of remembering these symptoms.

The patient in the question meets criteria for a manic episode. These symptoms have been ongoing for two weeks, she has decreased need for sleep, flight of ideas, increased goal-directed activity (purchasing clothes online), and increased talkativeness. The symptom of yelling at the sky implies psychosis, but we at least know that the symptoms are significantly impairing her home life—enough for her husband to bring her into the hospital.

 Are there any nonpsychiatric diagnoses possible?

Many diagnoses in the DSM-5, including bipolar and depressive disorders, have the disclaimer that they must not be caused by substance use or an underlying illness. In this scenario, we have no indication that there is an underlying nonpsychiatric explanation for the patient's symptoms.

 What diagnosis matches these symptoms?

The presence of a manic episode alone is enough to make the diagnosis of bipolar I disorder; however, the patient's history of depression strengthens this diagnosis because most patients with bipolar I disorder cycle between manic and depressive episodes.

This patient was originally diagnosed with major depressive disorder. This diagnosis requires at least one major depressive episode but cannot include manic episodes. Therefore, this patient's diagnosis is more accurately reflected by bipolar I disorder than major depressive disorder.

Bipolar II disorder has similar symptoms but includes hypomanic episodes instead of manic episodes.

Takeaways

Manic episodes are at least one week of elevated, expansive, or irritable mood with at least three of these symptoms: distractibility, insomnia, increased self-esteem or grandiosity, flight of ideas or racing thoughts, increased goal-directed or agitated motor activity, increased talkativeness or pressured speech, and high-risk behavior. They cannot be accounted for by a drug or underlying disorder.

Things to Watch Out For

As with many diagnoses in the DSM-5, a patient must fit the criteria of a disorder to be diagnosed with that disorder. Other related disorders, such as bipolar II disorder and major depressive disorder, have some shared symptomatology but do not meet the criteria for bipolar I disorder.

Related Questions

1. What is a depressive episode? What diagnoses may include a depressive episode?

2. What is a hypomanic episode? What diagnoses may include a hypomanic episode?

3. This patient's bipolar I disorder was not recognized until she started medication for a major depressive episode. Why is this commonly the case?

High-Yield Problem-Solving Guide questions continue on the next page. ▶ ▶ ▶

Key Concepts

Behavioral Sciences Chapter 8

Social processes
Bystander effect
Deindividuation
Peer pressure
Social facilitation

S Social Processes

Social actions and social phenomena impact a huge proportion of life for all of humanity, regardless of culture or location. Some of the more notable effects of social situations on the behavior of individuals have been evaluated and characterized. Using the definition of social action, assess the following series of situations, evaluate how behavioral response is altered by size of group, and identify the relevant social phenomenon.

1 What is a social action?

Action refers to any behavior performed by an individual in any circumstance. Social action refers to an action that accounts for the actions and reactions of other individuals. For instance, an action would be feeding oneself when hungry. That action would only be a social action if some aspect of the behavior, such as method of eating or choice of food, was selected based on the expectations or reactions of others. Not all actions are social, but all social actions must involve both an action and consideration of others or societal norms.

2 Situation 1: Tyler is attending a conference and notices a stranger fall down several stairs.

A stranger in peril would be most likely to evoke a social phenomenon known as the bystander effect. This effect occurs in social groups wherein individuals do not help a victim when others are present. It has been shown that the number of individuals in the room is inversely related to the likelihood and timeliness of the response to the victim. If fewer individuals were present, you would expect it to be more likely that Tyler would come to the aid of the victim.

Small Group: Tyler assists the stranger.

Large Group: Tyler does not respond.

3 Situation 2: Alex was attending an event and some of the other attendees became violent.

A group behaving in a way that the individuals involved would not typically behave is demonstrating the phenomenon of deindividuation. The group setting increases the anonymity of the behavior and can result in behavior that is inconsistent for the

Takeaways

The bystander effect describes the observed phenomenon that being in a group makes individuals less likely to come to the aid of a victim. The number of individuals in the group is inversely related to likelihood and timeliness of response. The effect is also influenced by cohesiveness of the group and the perceived danger of the situation.

Things to Watch Out For

Make sure you realize that the inverse relationship means that the more people there are in the group, the less likely it is that any one individual will respond, and that if an individual responds, the response time will generally be slower.

individual. Group cohesiveness and group size are directly related to the prevalence of deindividuation and lack of socially responsible behavior. With fewer individuals present or less group cohesiveness, you would expect it to be less likely that Alex would join in the violence.

Small Group: Alex does not participate in violent acts.

Large Group: Alex participates in violent acts.

 ## Situation 3: Adam had other plans, but his friends want to study and invite him to join.

An individual who is receiving social influence from a group of peers is most likely to experience the phenomenon of peer pressure. Peer pressure can be negative or positive, depending upon the situation. Adam's situation involves peers influencing him to behave in a more positive way, by encouraging him to study. This effect is dependent on the degree of familiarity with and perception of the peer group, and is not strongly impacted by group size. Adam's response would be dictated by how close he is with these friends and how highly he regards them. The greater his regard or familiarity, the more likely he is to study instead of play video games.

Small Group: Adam may or may not study.

Large Group: Adam may or may not study.

 ## Situation 4: Isaac knows a piece well, and plays in front of a large audience.

When engaging in behavior in front of a group rather than by oneself, social facilitation can impact performance. This effect is linked to increased likelihood for improved performance in the presence of others. This is more likely if the person feels a high sense of self-efficacy with the task at hand. The number of people present can slightly impact the level of arousal of the individual, but confidence and comfort with the task have a far larger effect. Regardless of group size, given that Isaac has been practicing for several weeks, his high self-efficacy should lead to an improved performance when in front of a group.

Small Group: Isaac's performance is improved compared to practice.

Large Group: Isaac's performance is improved compared to practice.

Related Questions

1. Children were observed on Halloween while trick-or-treating. Candy was left on the porch of a home with a note saying *Please take only one piece of candy*. It was observed that children trick-or-treating alone or with one friend or sibling were very likely to take only one piece of candy, but that children in groups were dramatically more likely to take handfuls of candy. What social process is at play, and what influence does the nature of the holiday have on the observed results?

2. An Olympic weightlifter is preparing for a competition and records his clean-and-jerk weights over a period of three months during his independent training sessions. He sees increases in his lifts as he progresses toward the competition, but plateaus in the final month. At the competition, he reaches a personal record, adding seven kilograms to his lift. What social process best describes the improved performance at the competition?

3. Max is a high school student who has typically shown mediocre performance. He transfers to a new school and befriends a group of high-achieving students. Max's grades in his first semester are much above his previous grades. What social process describes these results?

High-Yield Problem-Solving Guide questions continue on the next page. ▶ ▶ ▶

☰ Verbal and Nonverbal Communication

> Tom is on the interview trail for medical school. He wears a tailored black suit and a pin from his pre-medical honor society to every interview and makes references to the influential physician-scientists with whom he's done research. At one particular school, he arrives slightly late to the interview. *Pardon me for being late*, he says. *I couldn't find where to park.* What are some of the impression management strategies Tom is using during this interview day?

1 What is impression management?

Impression management refers to a person's attempts to influence how others perceive him or her. This is accomplished through the regulation or controlling of information in social interactions. There are a number of impression management strategies that are used in specific social situations.

2 What elements of the situation constitute impression management strategies?

There are two main elements of impression management in this story. First, Tom is wearing particular clothes—a tailored black suit and the pin from his pre-medical honor society—and creating a positive image by associating himself with influential physician-scientists. Second, he is providing a socially acceptable excuse for why he arrived late at an interview. Whether the excuse is actually true has some bearing on the label we can give to this impression management strategy.

3 Wearing specific clothes and referring to professional associations is what strategy?

The strategy of using props, appearance, emotional expression, or associations with others to create a positive image is called managing appearances. This impression management strategy usually requires a balance between presenting the authentic self—who Tom actually is—and the tactical self—who he markets himself to be when he adheres to others' expectations of him. While managing appearances can play a role in any social situation, we commonly think of its application in settings where an individual wishes to establish authority or to appear knowledgeable and trustworthy, especially in cases where a power dynamic exists.

4 What strategy is Tom using when he mentions not being able to park?

We are not given enough information to know whether Tom's claim that he could not find parking is true. Regardless, Tom is trying to make questionable behavior (being late for a medical school interview) acceptable through an excuse. This strategy is called aligning actions, and it is most commonly used when someone fails to live up to a particular expectation, such as being on time for an interview, scoring poorly on an exam, or missing a deadline.

Aligning actions is often associated with an external locus of control, in which someone considers personal successes or failures to be the result of factors outside of his or her control. Aligning actions also encourages others to make situational attributions about an individual's behavior, inferring that the causes of the behavior are features of the surroundings, rather than personal beliefs, attitudes, or personality characteristics.

Takeaways

Aligning actions is related to excuse-making and rationalization of questionable behavior. It may be associated with an external locus of control and may be used to encourage others to make situational—rather than dispositional—attributions.

Things to Watch Out For

Impression management techniques can overlap, and one may use multiple impression management techniques in the same social situation. Be prepared for questions that ask about mixing these strategies.

Related Questions

1. How does nonverbal communication add to a conversation?

2. What is alter-casting?

3. What is ingratiation?

Ⓢ Attribution Theory

You are chatting with your friend Carlton when a group walks by and one person casually mentions that they didn't get any studying done over the weekend for the upcoming Bio midterm. Carlton shakes his head and says, "Some people are just not made for pre-med courses. They procrastinate all the time and can't get things done." You point out to Carlton that he didn't get any studying done this weekend, either. He says, "Sure, but I studied all of last week and I'm doing great in class! Everyone was telling me I deserved a break, and I'm definitely going to study tonight!" Why did Carlton use a different type of attribution for himself as opposed to the stranger?

① Why does it matter who the perceiver is?

Attribution theory explains why and how individuals explain events the way they do. Attribution theory focuses specifically on the thought processes of the individuals who draw conclusions about the behaviors of others. These individuals are the perceivers—simply put, they perceive the actions of others (targets). The actual cause of the targets' behavior is not what is being described by attribution theory. When you try to identify an attribution, knowledge of the actual behavioral motivation is not necessary. Only the assessment of the perceiver constitutes an attribution.

② What attributions are made by Carlton?

There are two main categories of attributions: dispositional and situational. Dispositional, or internal, attributions are related to the features or characteristics of the person whose behavior is considered, including his or her beliefs, attitudes, and personality characteristics. In this case, when Carlton immediately assumed the stranger was procrastinating and "not made for pre-med courses," Carlton attributed the behavior to an internal trait of the stranger, believing the stranger did not complete his studying because of an overarching attitude of procrastination and inability to complete work. Carlton's explanation of the stranger's behavior is a dispositional attribution.

Situational attributions are external and based on the features of the surrounding environments, rather than long-term personality traits of the individual. When Carlton later considers his own behavior, Carlton cites his own ongoing solid performance, examples of peers encouraging him to take a break, and his own future study plans as external causes for his lack of studying. Carlton's explanation of his own behavior is a situational attribution.

 ### How does the fundamental attribution error apply to the given situation?

When examining the causes of others' behavior, there is a general bias toward making dispositional (rather than situational) attributions, which is known as the fundamental attribution error. In this case, Carlton's assumption was that the stranger was a habitual procrastinator. If Carlton had not made the fundamental attribution error, he might have considered other external reasons for the stranger's study habits, including those situational attributions Carlton made for his own lack of studying.

 ### If you also didn't study, how would Carlton most likely attribute your behavior?

The fundamental attribution error is more common in negative contexts and when dealing with others whom you do not know well. When making attributions for close friends, the fundamental attribution error is substantially less likely. Carlton would most likely attribute your behavior to situational causes, similar to his explanation for his own behavior.

Takeaways

We have a tendency to blame others. When inferring causes of others' behavior, we tend to make dispositional—rather than situational—attributions. When rationalizing our own mistakes or shortcomings, we tend to assume an external—rather than internal—locus of control.

Things to Watch Out For

Think about attribution theory and fundamental attribution error any time you see one person analyzing another person's behavior. Be on the lookout for assumptions that lead to dispositional attributions.

Related Questions

1. What theory describes the tendency to make dispositional attributions for behavior we see as intentional?

2. When Itai wins his event at a track meet, he says it is because he trained hard and performed his best. When he loses the following week, he says it is because he had a poor lane position and the ground was soft from the rain. What describes his views of success and failure?

3. Kristin is usually at the top of the class and gets As on all of her tests. You are surprised to hear that she got a D on her latest math test. What cue are you likely to use to explain her usual behavior? What type of attribution are you likely to make for this low math test score?

High-Yield Problem-Solving Guide questions continue on the next page. ▶ ▶ ▶

Key Concepts

Behavioral Sciences Chapter 11

Demographic structure

Fertility

Migration

Mortality

Urbanization

⊟ Demographics

> A hypothetical town currently has 1000 people. Of the 1000 people, 480 are women and 520 are men; 170 individuals self-describe as white, 280 as black, 230 as Asian, and 320 as Hispanic. The crude mortality rate is 15 for every 1000 persons per year. In one year, 45 children are born, 200 Hispanic individuals leave the town, and 150 Asian individuals enter the town. How has the population changed in terms of size and racial demographics? Which factors contributed to population growth or decline?

1 · What details given contribute to population size?

Population size is affected by reproduction, death, and net migration rates. The mortality rate for the town is 15 for every 1000 persons per year and the birth rate is 45 for every 1000 persons per year. The difference in the two provides the rate of natural increase. Net migration rate is the difference between immigration into and emigration from an area. In total, 150 people immigrate into this town and 200 people emigrate from it.

2 · Is the population size expected to rise or decline?

Because the birth rate is higher than the death rate, the population is expected to naturally increase. The rate of natural increase is the difference between these two, or:

$$\frac{45\text{ births}}{1000\ \frac{\text{persons}}{\text{year}}} - \frac{15\text{ deaths}}{1000\ \frac{\text{persons}}{\text{year}}} = \frac{+30\text{ persons}}{1000\ \frac{\text{persons}}{\text{year}}}$$

In addition, the net migration rate is the difference between the immigration and emigration rates, or:

$$\frac{150\text{ immigrants}}{1000\ \frac{\text{persons}}{\text{year}}} - \frac{200\text{ emigrants}}{1000\ \frac{\text{persons}}{\text{year}}} = \frac{-50\text{ persons}}{1000\ \frac{\text{persons}}{\text{year}}}$$

Adding these two rates together gives us the overall population growth rate:

$$\frac{30\text{ persons}}{1000\ \frac{\text{persons}}{\text{year}}} + \frac{-50\text{ persons}}{1000\ \frac{\text{persons}}{\text{year}}} = \frac{-20\text{ persons}}{1000\ \frac{\text{persons}}{\text{year}}}$$

Thus, this population is shrinking at a rate of 20 persons per 1000 persons per year, or 2%.

3 How have the demographics of the population changed?

The population was initially 17 percent white, 28 percent black, 23 percent Asian, and 32 percent Hispanic. The birth and death rates per racial demographic are not given; do not assume that these rates will be uniform across the population. While we cannot derive predictions about the changes in demographics from these rates, we can predict some qualitative changes from the migration statistics.

The percentage of Asians increased dramatically over the year, growing from 230 to 380 due to immigration. If we assume that all deaths were in this group, we know that the minimum population of Asians is $380 - 15 = 365$ individuals. If we assume that all births were in this group, we know that the maximum population of Asians is $380 + 45 = 425$ individuals. We can confidently say that this group will be the largest demographic in the population.

The Hispanic population, which was previously the largest, will decrease to between 105 and 165 individuals ($320 - 200 = 120$; $120 - 15 = 105$; $120 + 45 = 165$). Depending on the changes in the white population, either the Hispanic or the white population is now the smallest demographic in the population.

Related Questions

1. India has seen a large population growth in the past 50 years; however, the growth of Mumbai is significantly higher than in outlying areas. What migration phenomenon best explains this trend?

2. The recent U.S. trend to outsource work to other countries is an example of what social shift?

3. Russia has a rate of natural increase of –4.93. What does this mean in terms of birth and death rates?

Takeaways

There are three main ways for a population to grow: have more babies, have fewer deaths, or have more people move into the population than out of it. There are three main ways for a population to shrink: have fewer babies, have more deaths, or have more people move out of the population than into it.

Things to Watch Out For

Be careful with the sign convention in demographics. To determine the overall rate of change in the population, subtract the mortality rate from the birth rate and then add the result to the net migration rate.

Key Concepts

Behavioral Sciences Chapter 12

Race inequities in health

Gender inequities in health

Class inequities in health

Race, gender, and class inequities in healthcare

⊟ Disparities in Health and Healthcare

Researchers studied four groups: high-income females, high-income males, low-income females, and low-income males. Which group is expected to have the best health profile? How do they differ? How do access and usage of healthcare services differ among these groups?

1 How do health profiles differ by gender?

Statistics show that females have better health profiles than males. Life expectancy for women is higher than for men. Death rates due to heart disease, cancer, chronic respiratory disease, and diabetes are higher for men. Men are also more likely to die from accidents, homicide, and suicide. While men have higher mortality rates, women tend to have higher morbidity rates.

2 How do health profiles differ by class?

Low-income groups have far worse health profiles than high-income groups. Low-income groups are more likely to have poor health and to die younger. They are more likely to have life-shortening diseases and to die from homicide or suicide. In addition, infant mortality rates are higher for low-income groups.

3 Which group is expected to have the best health profile?

According to gender and income status, the high-income female group is expected to have the best health profile.

Takeaways

It is important to be familiar with both health and healthcare trends with respect to race, gender, and class. It is more useful to be aware of the trends than to memorize specific statistics.

Things to Watch Out For

While females are more likely than men to seek out healthcare services, they often experience longer delays or more difficulty in receiving care than men.

4 How does use and availability of healthcare differ among groups?

We have identified that women have better health profiles than men. Women are also more likely to utilize healthcare services. They are more likely to be insured and to have routine visits to primary care doctors, and they have a higher frequency of physician visits per year. However, women experience more difficulty with access and delays in healthcare services. Low-income groups have poorer healthcare profiles and poorer access to and utilization of healthcare services. The high-income female and high-income male groups have better access to healthcare and higher usage than both low-income groups. Females of both groups will utilize healthcare more often than their male counterparts.

Related Questions

1. How do the health profiles of white Americans, African Americans, and Asian Americans compare?

2. A low-income female from a small town goes to her local urgent care center. She knows all of the nurses and doctors because of the size of the town. A high-income male from out of town has more severe symptoms and has been at the urgent care center for five hours when the woman arrives, but is seen after her. What is the most likely cause for the discrepancy in care?

3. Rachel is a 35-year-old female with a body mass index of 21.3 from a low-income group. Stephanie is a 35-year-old female with a body mass index of 34.5 from a high-income group. Rachel visits her primary doctor for routine checkups more frequently and reports high satisfaction with care. Stephanie is less likely to visit the doctor and has changed primary care providers three times in the past five years. What best explains this discrepancy?

Solutions to Related Questions

1. Organization of the Human Nervous System

1. The hindbrain is located where the brain meets the spinal cord and includes the medulla oblongata, the pons, the cerebellum, and the reticular formation. The medulla oblongata is responsible for regulating vital functions such as breathing, heart rate, blood pressure, and digestion. The pons contains sensory and motor pathways between the cortex and the medulla. The cerebellum is at the top of the hindbrain and helps maintain posture, balance, and coordination of body movements. Damage to this area causes clumsiness, slurred speech, and loss of balance; note that these are similar to the impairments caused by alcohol, which largely affects the cerebellum. The reticular formation controls general arousal processes and alertness.

2. The hypothalamus is subdivided into three areas: lateral, ventromedial, and anterior. The lateral hypothalamus is referred to as the hunger center and contains special receptors thought to detect when the body needs more food or fluids. The ventromedial hypothalamus is identified as the satiety center and provides signals to stop eating. The anterior hypothalamus regulates sexual behavior, sleep, and body temperature.

3. The cerebrum is composed of the basal ganglia, limbic system, and cerebral cortex; it can also be divided between left and right hemispheres. The basal ganglia coordinate muscle movements as they receive information from the cortex and relay this information to the brain and spinal cord. The basal ganglia include the extrapyramidal motor system, which gathers information about body position and carries this information to the brain and spinal cord, helping to smoothen movements and steady posture. Damage to this area is associated with Parkinson's disease, which is characterized by jerky movements and uncontrolled resting tremors, among other symptoms. The basal ganglia are also believed to play a role in schizophrenia. Within the cerebrum are ventricles filled with cerebrospinal fluid that ultimately flows into the central canal in the middle of the spinal cord. Research has linked abnormally enlarged ventricles with symptoms often seen in schizophrenia, including social withdrawal, flat affect, and catatonic states. The functions of the limbic system and cerebral cortex are detailed in Step 2 of the main question.

2. Vision

1. The duplicity theory of vision states that the retina contains two types of photoreceptors: cones and rods. Cones come in three types and are used for color vision and to perceive fine details. Cones are most effective in bright light (daylight or artificial light). In reduced illumination, rods function best and allow perception only of an achromatic, lower-resolution image. Rods allow for night vision, when cones cannot function. Overall, there are many more rods than cones; however, the fovea at the center of the retina contains only cones. As one moves further away from the fovea, the proportion of rods increases and the proportion of cones decreases. Therefore, visual acuity is best in the fovea, and the fovea is most sensitive in normal daylight vision.

K

2. Feature detection theory states that we interpret objects by assessing specific characteristics, such as lines, shapes, or specific kinds of motion to identify something of importance *vs.* something of little value. Fishing is a good example of feature detection applied to animals: the person fishing attempts to mimic the features of something recognizable to the fish as food. Some cell types involved in feature detection are cones (for color), parvocellular cells (for shape and boundary detection), and magnocellular cells (for motion).

3. Parallel processing is the psychological counterpart to feature detection theory and refers to our analysis of different attributes of an object through separate pathways before integrating them. Parallel processing requires the interpretation of color, motion, shape, and depth as separate entities, which are then combined to create a cohesive view of the world.

3. Associative Learning

1. At first, this may seem like a paradox—decreasing behavior is the goal of punishment, after all. However, Emily can decrease the frequency of the negative behavior by reinforcing incompatible behaviors. For example, she might provide treats when the cat stays down off of the shelf or pet the cat when it is in a different room.

2. In terms of associative learning, a fetish is an example of a conditioned reinforcer. The fetish object has become paired with sexual gratification such that fixation on it and its use provide satisfaction without a specific sexual context. Once paired, the experienced gratification becomes a reward (positive reinforcer) that increases the use of the fetish object.

3. Ted has been conditioned to associate the songs played in the restaurant with the smell or taste of the food served there. The food is the unconditioned stimulus, the music is the conditioned stimulus, and the feeling of hunger is both the unconditioned and conditioned response. If Ted stops going to the restaurant, extinction of the conditioned response—a decreased feeling of hunger when hearing this music—should follow.

4. Language

1. Babies usually begin to babble in the first year of life; babbling peaks between 9 and 12 months. From 12 to 18 months, infants add about one word per month to their vocabularies. Around 18 months, there is an "explosion of language" in which the infant learns dozens of words and begins to use inflection and gestures to indicate different meanings for the same word. From 18 to 20 months, the infant also begins to use two-word sentences. From two to three years, a child begins to form three-word (or longer) sentences. By age five, children have largely mastered the rules of language.

2. According to the nativist (biological) theory of language acquisition, humans have an innate capacity for language. This capacity is sometimes referred to as the language acquisition device, a theoretical pathway in the brain that allows infants to absorb and process language rules. It is thought that this device is triggered by exposure to language. Thus, the nativist theory posits that children must be exposed to language during a critical period between the age of two and puberty in order to fully develop their linguistic abilities.

3. The learning (behaviorist) theory states that language acquisition is accomplished through operant conditioning. Caregivers repeat and reinforce the sounds that mimic their own spoken language. Thus, over time, the infant perceives that certain sounds are highly valued and are reliably reinforced, while others have little value and are not reinforced, with the result of shaping the child's acquisition of language.

5. Theories of Emotion

1. The Cannon–Bard theory, in contrast to the James–Lange theory of emotion, states that neither the physiological arousal nor the corresponding visceral stimulation for a particular emotion is distinct enough for the brain to label that emotion. Instead, this theory states that sensory information is received and sent to both the cortex and the sympathetic nervous system simultaneously by the thalamus. In other words, the physiological and cognitive components of emotion occur simultaneously (rather than in sequence) and result in the behavioral component of emotion.

2. According to the Schachter–Singer theory of emotion, both arousal and the labeling of arousal on the basis of environmental cues must take place in order for an emotion to be experienced. The subjective experience of emotion arises from the interaction between changes in physiological arousal and the cognitive interpretation of that arousal. In the absence of any clear emotion-provoking stimuli, the interpretation of physiological arousal depends on what is happening in the environment. In other words, physiological arousal could be labeled as anger, fear, or happiness, depending on environmental cues.

3. Darwin believed that emotions are products of evolution and that as a result, emotions and their corresponding expressions are universal. Because all humans evolved the same set of facial muscles, these muscles would show the same expression when communicating an emotion, regardless of differences in society or culture. Paul Ekman identified seven universal emotions: happiness, sadness, contempt, surprise, fear, disgust, and anger. Also, his research showed that each of these universal emotions comes with a specific set of facial cues, regardless of culture or society.

6. Stages of Moral Reasoning

1. As described, this child has successfully resolved the conflict of autonomy *vs.* shame and doubt. When resolved favorably, an individual feels able to exert control over the world and to exercise choice as well as self-restraint. The next conflict should be initiative *vs.* guilt, which normally occurs between ages 3 and 6. If resolved favorably, the child will feel a sense of purpose, can initiate activities, and can enjoy accomplishment.

2. Extreme tidiness and messiness, according to Freud, are two sides of the same coin: both indicate fixation in the anal stage of psychosexual development, which normally occurs between one and three years of age. The difference lies in which part of the unconscious is stronger. The tidy roommate's superego influences his behavior, whereas the messy roommate's id influences his behavior. Remember that the superego is focused on perfectionism, while the id is focused on satisfying primal, inborn urges.

3. Lacking a more knowledgeable other, this child would likely have difficulty accomplishing tasks that are within his zone of proximal development. This refers to tasks that a child cannot do on his or her own but can accomplish with the

assistance of this more knowledgeable other. Consistent failure to learn new tasks could lead to a feeling of ineffectiveness and helplessness, thereby likely causing low self-efficacy. This could, in turn, lead to a persistent external locus of control, or a feeling that one's successes and failures are due to circumstance, rather than personal characteristics and actions.

7. Bipolar and Depressive Disorders

1. Depressive episodes last at least two weeks and include at least five of the following: feeling down or **sad**, changes in **S**leep patterns, loss of **I**nterest, feelings of **G**uilt, loss of **E**nergy, difficulty **C**oncentrating, changes in weight or **A**ppetite, **P**sychomotor retardation or agitation (feeling or seeming slowed down or agitated), and recurrent thoughts of death or **S**uicide. These can be remembered with the mnemonic **sadness + SIG E. CAPS**. As with manic episodes, depressive episodes must not be accounted for by an underlying disorder. Major depressive disorder, bipolar I disorder, and bipolar II disorder may all include major depressive episodes.

2. Hypomanic episodes are less severe than manic episodes. While the diagnosis of a hypomanic episode also requires three symptoms of a manic episode (four if the mood is irritable only), they do not cause significant impairment to everyday life, cannot have psychotic symptoms, and need not last as long (at least four days). Bipolar II disorder is characterized by hypomanic episodes with or without depressive episodes.

3. Manic symptoms and depressive symptoms are, to an extent, at opposite ends of the same mood spectrum. Patients who actually have bipolar I disorder or bipolar II disorder may first present with a depressive episode. Without any history of mania (at the time), the person may be diagnosed with major depressive disorder. However, treatments of depression are aimed at activating the person and increasing energy and mood. In a patient who has an underlying bipolar disorder, the activation caused by the medication can actually lead to a manic episode, unmasking the bipolar disorder.

8. Social Processes

1. The process seen with these trick-or-treating children is deindividuation. When in groups, individuals are more likely to behave in ways that they would not if they were alone. The group setting increases the anonymity of the behavior and can result in behavior that is inconsistent with a person's typical behavior. During Halloween, children are dressed up in costumes, which increases anonymity, making them even more likely to behave in atypical ways.

2. Social facilitation is at play in this example. Individuals are more likely to exhibit enhanced performance in the presence of others—especially if they already feel a high sense of self-efficacy with the task at hand. The lifter completes his training sessions independently, but during competition, judges observe him and there may be a cheering crowd. Being watched enhances his ability to perform, resulting in a new personal record.

3. Peer pressure refers to social influence placed on an individual by a group of peers or others one perceives as equals. While peer pressure is often seen as a negative influence, it can also be positive. In Max's case, the fact that his new group of friends are high-achieving students places pressure on him to perform at the level of his peers. His desire to be socially accepted by his peer group results in behavior to meet the norm—in this case, higher grades in school.

9. Verbal and Nonverbal Communication

1. Nonverbal cues serve several functions in communication, but their primary purpose is often to express emotions. Ekman's seven universal emotions are a classic example of these emotion-disclosing facial expressions that do not differ across cultures. Nonverbal communication can also be used to convey attitudes; for example, smiling at someone while maintaining eye contact conveys *I like you*. One can communicate personality traits through nonverbal communication as well; for example, someone who is outgoing may use bold, broad hand gestures, an energetic tone of voice, and voice inflection to add to verbal communication. Nonverbal cues may be culture specific: maintenance of eye contact, an acceptable amount of personal space, and the meanings of different postures or hand signals can vary between groups. In these cases, successful communication is often contingent on knowledge of cultural symbols and display rules.

2. Alter-casting is the imposition of an identity onto another person. In this impression management strategy, one assigns a role to another person. This is often done subtly by implying that a desirable quality is associated with a given behavior. Examples include a friend saying *A real friend would . . .* or Kaplan saying *As a good MCAT student . . .* in our books.

3. Ingratiation is the use of flattery or conforming to expectations to win someone over. In this impression management strategy, one may blindly agree with another person, may compliment another person before asking for a favor, or may simply try to live up to "good boy, good girl" imagery.

10. Attribution Theory

1. Correspondent inference theory describes the tendency to make dispositional attributions when behavior is seen as motivated or intentional as opposed to accidental. We tend to assume that unexpected actions are representative of an individual's personality or motives; thus, in cases when an individual unexpectedly performs a behavior that helps or hurts us, we assume that this is reflective of the person (dispositional attribution) rather than of circumstance (situational attribution).

2. Self-serving attributional bias is the tendency to attribute successes to dispositional factors and failures to situational factors. Itai sees his success as a result of hard work and personal achievement. On the other hand, he sees his failure as a result not of his own actions, but rather of lane position and track conditions. We could also say that successes are explained with an internal locus of control, whereas failures are explained with an external locus of control. It is noteworthy that in some cases of depression, self-serving attributional bias becomes reversed: the individual attributes success to situational factors (*I got lucky this time*) and failure to personal factors (*It was all my fault*).

3. When making attributions, we consider a person's behavior over time. Because Kristin has always gotten As, we expect her to continue to get As. These are consistency cues, which refer to having consistent behavior over time, as well as distinctiveness cues, which refer to having similar behavior in similar situations. On the latest test, Kristin got a D, which is not consistent with her past behavior. In this case, we are likely to attribute the behavior to situational as opposed to dispositional factors because we have familiarity with Kristin's "normal" behavior.

11. Demographics

1. While birth and death rates likely play a role in the disproportionate growth of Mumbai in comparison to outlying areas of the city, urbanization is also likely one of the main drivers of population growth. Urbanization is the migration of large groups of people to densely populated urban areas, creating cities. In fact, analysis of the outlying areas may demonstrate that these towns are actually decreasing in size as large portions of their populations move into the city for economic or social opportunities.

2. Outsourcing is an example of globalization. Globalization is defined as the increase in internationally connected systems and integration, with the tapping of foreign labor markets and increased availability of goods and services.

3. The rate of natural increase is the crude birth rate minus the crude death rate. Having a negative rate of natural increase translates to the death rate being higher than the birth rate in a population. This means that, assuming no immigration or emigration, the population will shrink at a rate of 4.93 percent per year.

12. Disparities in Health and Healthcare

1. Asian Americans have the best health profiles of the group, followed by white Americans. African Americans have the worst health profiles of the three. Specifically, Asian Americans have lower mortality rates associated with cancer, heart disease, and diabetes and lower infant mortality rates than other populations. African Americans, on the other hand, have higher mortality rates linked to cancer, heart disease, diabetes, drug and alcohol use, and HIV/AIDS and higher infant mortality rates than other populations.

2. It has been shown that preferential treatment often arises from in-group associations. The female in this story is part of the in-group because she knows everyone in the town—most importantly, the doctors and nurses who will be involved in her healthcare. The male, on the other hand, is part of an out-group, and thus, despite having more severe symptoms and spending a longer time in the waiting room or triage, still waits longer to receive healthcare services.

3. While it is expected that high-income females are more likely to visit the doctor and have better access to healthcare, obesity is potentially a factor in this case. Stephanie's body mass index puts her in the obese range (over 30). Obesity bias is an identified issue in healthcare. Obese patients are more likely to change doctors, and they report lower levels of trust in their primary care physicians. Doctors are also less likely to recommend effective weight-loss programs to obese patients, sometimes on the basis of the flawed assumption that obese patients lack the willpower to effectively lose weight. Obese patients are also less likely to be offered quality preventative care and screenings.

Biochemistry

Ε Elements of Peptide Structure

Creutzfeldt–Jakob disease is a neurodegenerative disorder caused by a prion protein. This prion protein induces a change in secondary protein structure from α-helices to β-pleated sheets. What are possible biochemical consequences of this change in secondary protein structure?

1 What are the characteristics of α-helices and β-pleated sheets?

α-Helices are rodlike structures in which the peptide chain coils clockwise about a central axis. The helix is stabilized by intramolecular hydrogen bonding that occurs between carboxyl oxygens and amino hydrogens located four residues away from each other. Typically, amino acid side chains point away from the helix's core, allowing for interaction with the cellular environment.

In β-pleated sheets, the peptide chains form rows that are held together by intramolecular hydrogen bonding that occurs between the carboxyl oxygen on one peptide chain and the amino hydrogen on another. Pleating maximizes hydrogen bonding in the structure.

2 How does structure impact function?

The structural properties of a peptide affect its function. After analyzing the structure of the protein, determine how each feature may affect the chemical properties of the protein. α-Helical structure involves extension of hydrophilic side chains away from the core of the protein toward the aqueous environment of the cell. β-Pleated sheets are less likely to assume this sort of confirmation; however, one side of the β-pleated sheet may contain hydrophobic side chains, while the other may be hydrophilic.

If we wanted to determine how structure affects protein degradation, we would need to compare the ease with which enzymes could access the amide bonds in each conformation. α-Helices are strandlike structures, while β-pleated sheets are more flat. An α-helix is more easily degraded because its peptide bonds are more accessible to *peptidases*—enzymes that cleave peptide bonds. On the other hand, β-pleated sheets are more stable and expose far fewer residues to the cellular environment.

How might changing a protein from α-helix to β-pleated sheet affect the organism?

Prions convert proteins from a more soluble, easily degraded α-helical conformation to a less soluble, more difficult to degrade β-pleated sheet conformation. Insoluble proteins that cannot be degraded will eventually build up and form plaques, which can lead to loss of cell function and ultimately cell death. As cells die, aberrant proteins may be left behind, resulting in pockets of protein within tissues, and complete loss of function of cells, tissues, and even whole organs.

Related Questions

1. A point mutation changes a cysteine residue to an alanine residue. How might this affect protein structure?

2. A protein is treated with a 6 *M* solution of hydrochloric acid. What levels of protein structure are most likely disrupted by treatment with this solution?

3. A point mutation causes a single leucine residue to be substituted for an isoleucine residue in the transmembrane section of a G protein-coupled receptor. How might this change affect overall protein structure and function?

Takeaways

α-Helices create proteins that are more strandlike, more soluble, and easier to degrade. β-Pleated sheets create proteins that are flatter, less soluble, and more stable (harder to degrade).

Things to Watch Out For

Many questions on the MCAT integrate knowledge from multiple science areas. While the focus in this question was on the biochemistry of prion diseases, other questions could focus on the fact that they can be genetically inherited or can be spread as an infectious disease.

§ Enzyme Kinetics

Enzyme activity is often graphed using a Michaelis–Menten plot. However, enzyme activity may also be graphed using an alternative format, the Lineweaver–Burk plot. HMG-CoA reductase is the rate-limiting enzyme of cholesterol synthesis. Statins, such as atorvastatin (Lipitor), simvastatin (Zocor), and rosuvastatin (Crestor) are routinely used to treat patients with high cholesterol because they are competitive inhibitors of HMG-CoA reductase. How would use of a statin affect the Lineweaver–Burk plot?

1 **What is the Michaelis–Menten plot's basic shape? How does it display K_M and V_{max}?**

Michaelis–Menten Plot

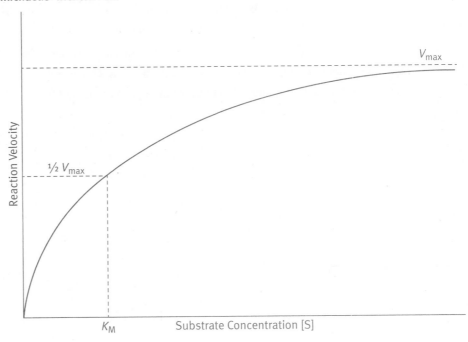

The Michaelis–Menten saturation curve exhibits asymptotic behavior. As substrate concentration increases, the reaction rate increases as well, up to an asymptotic maximum, designated V_{max}. The Michaelis constant, K_M, is defined as the substrate concentration at which enzyme velocity equals half of V_{max}. K_M is found on a Michaelis–Menten plot by finding the one-half V_{max} value on the y-axis, tracing that value over to the Michaelis–Menten curve, and then tracing down to the x-axis.

2 How does a Lineweaver–Burk plot display K_M and V_{max}?

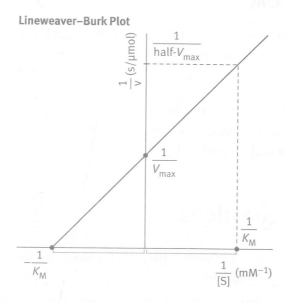

Lineweaver–Burk plots measure reaction rate as a function of the concentration of substrate available to bind to the enzyme. The y-axis represents the reciprocal of the velocity or reaction rate, $1/V$, and the x-axis represents the reciprocal of substrate concentration, $1/[S]$. The y-intercept is the reciprocal of the enzyme velocity when $1/[S] = 0$. If $1/[S] = 0$, then $[S] = \infty$. Thus, the y-intercept corresponds to $1/V_{max}$. The x-intercept in a Lineweaver–Burk plot corresponds to $-1/K_M$.

3 Why does a competitive inhibitor increase an enzyme's K_M while leaving its V_{max} unchanged?

Competitive inhibitors compete with the substrate for binding to the active site of an enzyme. The presence of these inhibitors results in fewer active sites available to act on the substrate, resulting in decreased enzyme activity. Competitive inhibition can be overcome by increasing the substrate concentration. Excess substrate can outcompete the competitive inhibitor, allowing the enzyme to regain V_{max}. Therefore, while it may take more substrate to reach V_{max}, the value of V_{max} itself does not change. As addition of more substrate is necessary to reach the same final maximum, the value of K_M for the enzyme will increase in the presence of competitive inhibitor. K_M can be thought of as a measure of the affinity between an enzyme and its substrate; a higher K_M indicates decreased affinity between an enzyme and its substrate.

Takeaways

Competitive inhibitors occupy the active site of an enzyme, but the binding of the inhibitor can be reversed if the substrate concentration is high enough. Competitive inhibition does not change V_{max}, but does increase K_M (the substrate concentration needed to achieve half-maximal velocity).

Things to Watch Out For

K_M can be easily misinterpreted. Because it is the concentration at which half-maximal velocity is reached, an increase in K_M can indicate an inefficient enzyme, the presence of an inhibitor, or a decrease in the affinity of the enzyme for its substrate.

 A competitive inhibitor will have the effect of increasing the slope of a Lineweaver–Burk plot. Why?

In the presence of a competitive inhibitor, V_{max} is unchanged. The y-intercept of the Lineweaver–Burk plot corresponds to $1/V_{max}$, so there is no change expected in the y-intercept.

In the presence of a competitive inhibitor, K_M increases. The x-intercept of the Lineweaver–Burk plot corresponds to $-1/K_M$. Therefore, an increase in K_M indicates a decrease in the magnitude of $-1/K_M$, resulting in the x-intercept moving closer to the origin.

Related Questions

1. How would the Lineweaver–Burk plot change in the presence of a noncompetitive inhibitor?

2. How would the Lineweaver–Burk plot change in the presence of an uncompetitive inhibitor?

3. If the information in the HMG-CoA reductase experiment had been represented as a Michaelis–Menten plot instead of a Lineweaver–Burk plot, what changes would be observed in the presence of a competitive inhibitor?

High-Yield Problem-Solving Guide questions continue on the next page. ▶ ▶ ▶

⑤ Amino Acid Electrophoresis

A biochemist is trying to separate glycine, glutamic acid, and lysine given the following information:

	pK_a, COOH group	pK_a, NH^{3+} group	pK_a, R group
Glycine	2.34	9.60	–
Glutamic acid	2.19	9.67	4.25
Lysine	2.18	8.95	10.53

The mixture of amino acids is loaded onto the center of a polyacrylamide gel. The gel has a pH of 6. An electric potential difference of 220 V is applied for forty-five minutes, and the three points below indicate the three final locations of the amino acids within the gel. Which point corresponds to which amino acid?

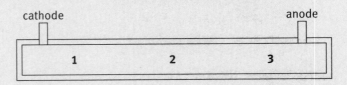

① **What is the conjugate base:conjugate acid ratio of its two functional groups at pH 6?**

COO⁻ : COOH ratio

$$pH = pK_a + \log\left(\frac{[COO^-]}{[COOH]}\right),$$

$$6 = 2.34 + \log\left(\frac{[COO^-]}{[COOH]}\right),$$

$$\sim 4 = \log\left(\frac{[COO^-]}{[COOH]}\right),$$

therefore, [COO⁻] : [COOH] = 10⁴:1.

NH₂ : NH₃⁺ ratio

$$pH = pK_a + \log\left(\frac{[NH_2]}{[NH_3^+]}\right),$$

$$6 = 9.60 + \log\left(\frac{[NH_2]}{[NH_3^+]}\right),$$

$$\sim -4 = \log\left(\frac{[NH_2]}{[NH_3^+]}\right),$$

Therefore, [NH₂] : [NH₃⁺] = 1:10⁴.

Glycine's isoelectric point is 6. The Henderson–Hasselbach equation is used to determine the ratio of conjugate base to conjugate acid when the pH of the solution and the pK_a of the group of interest are known.

Conclusion: When the pH is lower than the pK_a of a functional group, the acid form of the group predominates, and vice versa.

2 In what proportions do glycine's three ionizable states exist at its pI?

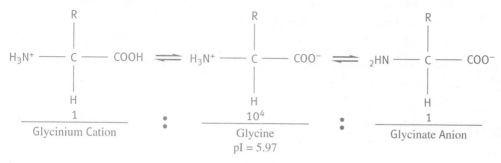

Glycinium Cation $\vdots$ Glycine pI = 5.97 $\vdots$ Glycinate Anion

Based on the previous calculation, at pH 6 the ratios of $NH_2:NH_3^+$ and of $COOH:COO^-$ are both $1:10^4$. Therefore, both the NH_3^+ and COO^- functionalities predominate at this pH, and the charge balance between these functional groups implies that the neutral form of glycine predominates at this pH. This observation is consistent with the fact that glycine's isoelectric point is approximately 6 (5.97, exact to two significant figures). So, a given glycine molecule most likely exists in its zwitterion form at this pH.

3 What charge do glutamic acid and lysine carry at a pH of 6?

At a pH of 6, all carboxylic acid groups will most likely be deprotonated (COO^-) because their pK_a values are all significantly lower than 6. All amino groups will most likely be protonated (NH_3^+) given that their pK_a values are significantly higher than 6. Drawing the charges makes it evident that at this pH, glutamic acid will carry a net negative charge and lysine will carry a net positive charge.

$$
\left[\begin{array}{c} RCOO^- \\ pK_a = 4.25 \\ {}^+H_3N - C - COO^- \\ pK_a = 9.67 \qquad pK_a = 2.19 \\ H \end{array} \right]^{-1}
\left[\begin{array}{c} RNH_3^+ \\ pK_a = 10.53 \\ {}^+H_3N - C - COO^- \\ pK_a = 8.95 \qquad pK_a = 2.18 \\ H \end{array} \right]^{+1}
$$

Glutamic Acid
pI = 3.22

Lysine
pI = 9.74

Takeaways

To solve isoelectric focusing questions, determine the pI of the samples being separated. Then, if the gel has a pH gradient, determine its orientation in the gel. As a rule of thumb, regardless of whether there is a pH gradient or not, anions are attracted to the anode which is the acidic part of the gel.

Things to Watch Out For

Be careful with problems that ask you to separate proteins. Remember that the pI of a protein cannot be determined simply by averaging the pI values of the individual amino acids. If the isoelectric point of a protein is not given, then another method must be used to separate the proteins.

 ## To which point does each amino acid migrate?

Electrophoresis gels are always run as electrolytic cells. This means that the negative terminal of the gel is connected to the cathode of the battery, and the positive terminal of the gel is connected to the anode of the battery. The cathode attracts cations, while the anode attracts anions. The gel has a pH of 6. At this pH, glycine will carry a net neutral charge, glutamic acid will be negative, and lysine will be positive. Therefore, glycine will remain at point 2 where the samples are loaded, lysine will migrate toward the cathode (point 1), and glutamic acid toward the anode (point 3).

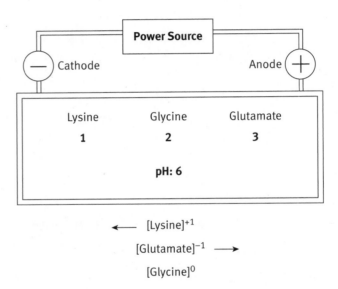

Related Questions

1. If a polypeptide with a pI of 6.7 is subjected to electrophoresis at pH 5, will the segment move toward the anode or the cathode?

2. If the gel in the original question had a pH gradient, would the time electrophoresis had run have mattered?

High-Yield Problem-Solving Guide questions continue on the next page. ▶ ▶ ▶

Key Concepts

Biochemistry Chapter 4

Fischer projections
Carbohydrate structure
Carbohydrate terminology
Stereoisomers
Chiral centers

E Isomerism in Carbohydrates

What is the relationship between the two structures presented below?

$$
\begin{array}{c}
\text{CHO} \\
\text{H} \!-\!\!-\!\!-\! \text{OH} \\
\text{H} \!-\!\!-\!\!-\! \text{OH} \\
\text{H} \!-\!\!-\!\!-\! \text{OH} \\
\text{CH}_2\text{OH}
\end{array}
\qquad
\begin{array}{c}
\text{CHO} \\
\text{HO} \!-\!\!-\!\!-\! \text{H} \\
\text{H} \!-\!\!-\!\!-\! \text{OH} \\
\text{H} \!-\!\!-\!\!-\! \text{OH} \\
\text{CH}_2\text{OH}
\end{array}
$$

1 What structural similarities exist between these molecules?

Start by carefully analyzing each carbohydrate molecule, taking note of similarities between the structures. Both structures are aldoses that contain five carbon atoms and are therefore termed aldopentoses. The highest-numbered chiral carbon (located farthest from the carbonyl group) contains a hydroxyl group (–OH) pointing to the right in the Fischer projection, making the configuration D for both sugars. Remember that while D- and L-isomers of the same sugar are enantiomers, this nomenclature is relative to the stereochemistry of the highest-numbered chiral carbon in glyceraldehyde, and has no automatic correlation to the direction of rotation of plane-polarized light.

2 What structural differences exist between these compounds?

The only difference between the structures is the configuration of C-2. In the first structure, the –OH group points to the right, giving this carbon an (R) configuration, while the –OH group points to the left in the second structure, giving this carbon an (S) configuration.

3 What terminology should be used to describe this difference?

When describing the isomerism of sugars, there are a few words we could use to describe differences between stereoisomers. We already discussed enantiomerism in Step 1;

enantiomers differ in configuration at all chiral carbons and are nonsuperimposable mirror images of each other. These two sugars are not enantiomers because they differ at only one of their three chiral carbons.

Isomers that differ at some—but not all—chiral centers are considered diastereomers. While this term would be an appropriate choice to describe the differences between these sugars, the MCAT tends to choose the most specific term possible to describe isomers. Epimers are diastereomers that differ at exactly one chiral carbon and can be named based on this carbon; these two molecules could be called C-2 epimers of each other.

Finally, anomers are a subtype of epimers in which the chiral center that differs between the two sugars is the anomeric carbon (the chiral center that is created by ring closure of the sugar, in which one of the hydroxyl groups attacks the carbonyl carbon in a nucleophilic addition reaction). These molecules cannot be considered anomers of each other because they are in their straight-chain forms.

Thus, these molecules are epimers, a subtype of diastereomers. These structures are D-ribose (on the left) and D-arabinose (on the right).

Takeaways

Carbohydrate isomerism utilizes unique terminology. On Test Day, you must be familiar with the specific vocabulary associated with carbohydrates (aldose, ketose, furanose, pyranose, epimer, anomer, and so on).

Things to Watch Out For

Take the time to identify the distinguishing characteristics of each sugar structure. Checking the relative (*not* absolute) configurations of each chiral carbon can prevent you from missing important details.

Related Questions

1. Mutarotation is the interconversion between the α- and β-cyclic forms of a sugar. In solution, the α and β forms are in equilibrium. What is the term that describes the relationship between these two sugars?

2. Cellulose is created from β-D-glucopyranose, while glycogen is created from α-D-glucopyranose. Human digestive enzymes cannot break down cellulose while glycogen is readily degraded to release glucose. What accounts for this difference?

3. Identify the pairs of epimers and their relationships in the set of molecules below:

D-fructose   D-glucose   D-galactose   D-mannose

Key Concepts

Biochemistry Chapter 5

Saponification
Triacylglycerols
Enzymatic cleavage
Anions and cations

E Lipid Saponification

Inflammation of the pancreas, or pancreatitis, occurs due to premature activation of digestive enzymes in the pancreas. Hypocalcemia, or lowered blood calcium concentration, may result from lipase hydrolysis of triacylglycerols in areas surrounding the pancreas. What is the most likely mechanism by which this hypocalcemia occurs?

1 How could you simplify this question?

Some questions on Test Day may appear very complex, and it may even seem that there is not enough information provided. However, simplifying and rewording the question may turn a difficult-appearing question into a much simpler one. The question stem states that *lipase* cleaves triacylglycerols, resulting in a lowered calcium concentration in the blood. A simpler form of this question might be: *What are the products of triacylglycerol hydrolysis and why do these products lower blood calcium?*

2 What are the products of triacylglycerol hydrolysis by lipase?

Enzyme names are one of the many areas of biochemistry in which nomenclature is helpful. Enzymes are generally named after their substrates or main functions. In this case, the name *lipase* implies that the substrate is fat (*lip–*); *–ase* is a suffix identifying the entity as an enzyme.

Human *pancreatic lipase* digests triacylglycerols to form two free fatty acid molecules and one molecule of 2-monoacylglycerol. Like the pancreatic proteases (*trypsin*, *chymotrypsin*, and *carboxypeptidases A* and *B*), pancreatic lipase is secreted as a zymogen—a proenzyme that must be activated to carry out its function. This is mostly a protective function; digestive enzymes cannot differentiate between self and nonself, thus pancreatic lipase is only activated once it reaches the duodenum, preventing autodigestion of the pancreas. Not all of these details are absolutely necessary to answer the question. The most important thing to note about lipase is its main function: that it causes the release of free fatty acids from dietary triacylglycerols.

③ What product is capable of binding to calcium?

Calcium is a Group IIA (Group 2) alkaline earth metal; as such, it normally exists in compounds or as a +2 cation. If calcium is being pulled from the bloodstream, it is most likely forming an ionic bond with one of the products of saponification. Therefore, let's determine if one of the products is negatively charged.

Water is required in the hydrolysis of triacylglycerols by pancreatic lipase. The water can protonate the hydroxyl groups of glycerol once the free fatty acids are released, as well as the hydroxyl group within the carboxyl moiety of the free fatty acid itself. However, this reaction takes place within the alkaline environment of the pancreas (pH ≈ 8.5), which normally releases bicarbonate to neutralize the acidic chyme dumped into the duodenum by the stomach. Thus, this reaction does not take place in an environment that favors the protonation of these compounds. The stronger acid between glycerol and the free fatty acids will be more easily deprotonated, and will exist in a negatively charged form.

Fatty acids contain a carboxylic acid group, the pK_a of which is usually around 4. Glycerol contains alcohols, the pK_a values of which are usually around 17. While it is not necessary to have memorized the various pK_a values of the different functional groups, you should recognize that a carboxylic acid is generally much more acidic than an alcohol.

Therefore, we would expect that most of the free fatty acid molecules would exist in a negatively charged state. Calcium ions are positively charged and will react with fatty acids to form a chalky white substance. This is an example of saponification, which is essentially the hydrolysis of triacylglycerols followed by binding of a cation to the free fatty acids.

Takeaways

Saponification does not always involve sodium hydroxide, although NaOH and KOH are frequently used for this reaction. Saponification is a general term for soap-making, in which an amphipathic carboxylate group is bound to cations after hydrolyzing a lipid ester to form a polyol and free fatty acids.

Things to Watch Out For

Many questions are interdisciplinary and require diverse knowledge in multiple areas. In this particular example, it is necessary to understand biochemistry (lipid structure), biology (the digestive system), general chemistry (group trends), and organic chemistry (functional groups' pK_a values). Expect that the MCAT will ask some questions that require you to bring together multiple content areas.

Related Questions

1. Common household soap is produced by reacting a fat or oil with a base to form sodium or potassium salts of long-chain fatty acids. When used in hard water, which contains high concentrations of mineral salts, a precipitate may form that is difficult to remove. What is this precipitate, and what other conditions might reduce the effectiveness of soap?

2. Adipocere, or grave wax, will form on a dead body buried in cold, humid, and low-oxygen conditions. What is likely to account for the formation of adipocere?

3. What would result following treatment of a triacylglycerol with a strong acid rather than a base? Could this result in saponification?

⑤ Central Dogma of Genetics

Werner syndrome is a form of adult onset progeria caused by a mutation to a single protein, the WRN protein, a helicase used in single- and double-strand break repair. There are many types of mutations that can potentially affect the WRN protein. To characterize the specific type of mutation in a given subpopulation of Werner patients, researchers run the following tests:

- A Northern blot reveals that while Werner patients do produce mRNA transcripts of the WRN gene, those transcripts are found to run at a lower than normal molecular weight.
- WRN protein is extracted from an affected patient and is found to function normally.
- A Western blot finds lower than normal concentrations of WRN protein in affected patient cells.

Using these data, what is the impact of the WRN mutation affecting this patient subpopulation?

① What is the typical function of helicase class proteins?

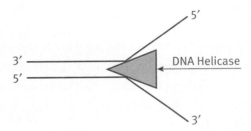

Helicases are a class of enzymes that are particularly important for DNA replication. The major function of helicase class enzymes is stabilization and separation of the DNA double helix. In Werner syndrome, the role of the WRN helicase most probably involves opening and stabilizing the DNA double helix at necessary repair sites during the synthesis phase of the cell cycle.

Key Concepts

Biochemistry Chapter 6
DNA
mRNA
tRNA
Transcription
Translation

2 What type(s) of mutations are implicated by the Northern blot data?

If there are mRNA transcripts for the WRN protein present, then the gene is being transcribed. Post-transcription, a typical mRNA molecule has a 5′ cap, a poly-A tail, 5′ and 3′ untranslated regions (UTRs), and a coding sequence. Given that the gene in question is a helicase and would be used in the nucleus, you can also infer that the coding region must contain a nuclear localization sequence (NLS), which would allow the protein to be taken into the nucleus post-translation. The knowledge that the gene is being transcribed allows you to eliminate the possibility of the removal of a start codon or the possibility of issues with transcription itself. However, the Northern blot data reveals that the present mRNA transcript has a lower molecular weight than normal, indicating that it is a shortened transcript. These mutations could affect any portion of the mRNA transcript, which includes both the coding sequence and the 5′ and 3′ untranslated regions (UTRs). Within the coding sequence, either the functional code for the protein or the nuclear localization sequence (NLS) could be affected. However, the 5′ cap and Poly-A tail are added after transcription to stabilize the mRNA and allow it to leave the nucleus, and would not be affected by a mutation that affected the DNA of the WRN helicase gene, so they are not the site of mutation.

Cap	5′ UTR	Coding Sequence	3′ UTR	Poly-A Tail

3 How is this mutation able to cause visible symptoms if the protein is functional *in vitro?*

If the WRN protein is functional in Werner syndrome patients, the mutation must generate symptoms via a mechanism aside from protein function. The other potential variable affected by this mutation is protein concentration. Werner patients experiencing symptoms must be lacking sufficient quantities of the protein itself due to the mutation and its impact on translation. This implicates two possible regions on the gene: the UTRs, which stabilize the transcript and facilitate translation, or the NLS, which allows the translated protein to enter the nucleus.

4 Given the Western blot data, what must this mutation be impacting?

Blot for WRN Helicase:

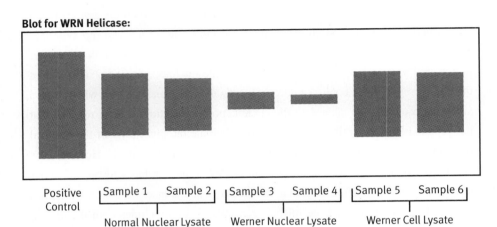

Positive Control | Sample 1 · Sample 2 | Sample 3 · Sample 4 | Sample 5 · Sample 6
Normal Nuclear Lysate | Werner Nuclear Lysate | Werner Cell Lysate

The Western blot contains data for nuclear lysate of a non-affected cell, nuclear lysate of an affected cell, and whole cell lysate of an affected cell. Compared to the normal cell, whole cell WRN helicase concentration looks relatively normal, but nuclear lysate reveals substantially diminished protein concentration. If protein concentration in the cytosol is normal in this patient population, translation itself is functioning as expected and the mutation is not in the 5′ nor 3′ UTR. A decreased nuclear lysate for the WRN helicase patients confirms that the protein is unable to enter the nucleus post-translation, confirming a mutation to the NLS.

Takeaways

The central dogma of biology is that DNA → RNA → Protein. If there is a mutation present in the DNA, it may be expressed via this pathway and eventually result in a change to the resultant protein.

Things to Watch Out For

Though every enzyme catalyzes just one process, that process may be involved in multiple events. Each enzyme involved in DNA replication can also potentially be involved in DNA repair. If DNA replication does not occur properly, mutations accumulate within the genetic material.

5 Why might this mutation have this functional impact?

Even though the protein is being produced normally, it is unable to be transported to the region in which it is used. High levels of WRN protein outside of the nucleus cannot act to assist with repair functions inside the nucleus. Thus, even though the protein is being made, it is not able to function in the patient population.

WRN helicase is involved in repair machinery. The mutation in the DNA for this gene leads to a transcript that is unable to form a translated protein that can be used in the nucleus. With decreased concentration of the WRN protein available in the nucleus, Werner syndrome most likely functionally results in decreased rates of successful DNA repair. This failure would lead to mutation accumulation that could have a myriad number of deleterious effects, most likely including increased rates of mutation-related cellular diseases such as cancer.

Related Questions

1. Polymerase chain reaction, or PCR, may be performed to amplify small samples of DNA for analysis. In PCR, a sample is denatured with heat, replicated, and then cooled to reanneal the strands. The strands undergo this process several times to rapidly increase the amount of sample available for biochemical testing. The DNA polymerase used for this process is *not* generally human DNA polymerase, but that of *T. aquaticus*, a bacterium. Why is human DNA polymerase NOT used for PCR?

2. Which strand is more prone to mutations, the leading strand or the lagging strand?

3. A molecule of DNA is replicated using three cycles of PCR in the presence of ^{15}N-labeled adenine. How many of the newly formed DNA molecules will contain at least one unlabeled strand?

E Operons

Prokaryotes often use operons to control gene expression. The *lac* operon, present in prokaryotes, has been studied extensively. In the presence of lactose and the absence of glucose, how are the appropriate genes induced? What happens when lactose levels fall and glucose levels rise?

1 What is the role of the *lac* operon?

Operons are functional units found in bacterial chromosomes and include structural genes that can code for cellular proteins such as enzymes. In addition, a typical operon also contains a promoter, an operator region, and a regulatory gene. The promoter is the site at which *RNA polymerase* binds. The operator serves as the binding site for the repressor protein encoded by the regulatory gene. All of the genes within the operon are controlled by this unique mechanism. When the repressor binds to the operator region, the polymerase is blocked and transcription does not occur.

The *lac* operon is a prototypical inducible system. The genes in the *lac* operon regulate enzymes needed to break down lactose in prokaryotes. Gene transcription is a significant energy cost for an organism. In order to prevent wasteful use of energy, inducible systems allow for the transcription of genes only when they are beneficial in the current environment. The *lac* operon, in particular, allows for transcription of genes essential to the digestion of lactose when glucose is not present.

2 What is the function of the *lac* operon?

As mentioned above, the *lac* operon is an inducible system that encodes genes used in the digestion of lactose. The repressor protein is normally bound to the operator region, preventing transcription. When the disaccharide lactose enters the cell, it binds to the repressor. This binding is considered a form of allosteric regulation. The binding of lactose causes a conformational change, resulting in dissociation of the repressor from the operator DNA sequence, making those genes available for transcription. However, this is not the end of the story—there has to be something that attracts RNA polymerase to the site to transcribe those genes.

What is the process by which these genes are transcribed?

Once the repressor has been unbound from the operator region, the normal process of gene transcription can occur. However, it is important to recognize that transcription of the genes for lactose digestion require both the presence of lactose *and* the absence of glucose.

If the promoter sequence varies from the consensus sequence, then RNA polymerase will not bind tightly without help. The solution to this problem is the catabolite activator protein (CAP), which is present when glucose levels are low. In the absence of glucose, *adenylate cyclase* is also active and produces cAMP. CAP binds cAMP, and this complex binds to the CAP-binding site located just before the promoter region. Once bound, the complex helps recruit RNA polymerase, which can then start transcribing the genes for lactose digestion.

It is not necessary to know all of these details to answer this question; the important point to note is that the transcription of the genes for lactose digestion requires not only that lactose is *present*, but also that glucose is *absent*.

What happens when lactose levels fall and glucose levels rise?

As lactose levels fall, lactose will dissociate from the repressor, which then binds to the operator region. The presence or absence of glucose also affects the intracellular concentration of cyclic AMP (cAMP) because catabolites of glucose will inhibit adenylate cyclase and inhibit the production of cAMP. Without cAMP, the cell cannot transcribe the genes needed to metabolize lactose, even when lactose is present. This results in the preferential use of glucose. Both of these conditions help to decrease the production of the enzymes required for lactose digestion.

Related Questions

1. What happens to the *lac* operon in the presence of both lactose and glucose?

2. What is the difference between an inducible system and a repressible system?

3. The *trp* operon governs the synthesis of tryptophan by prokaryotes. In a tryptophan-poor medium, what is the process by which the genes for tryptophan synthesis can be turned on?

Takeaways

Operons are complex systems used to control gene transcription. Inducible systems allow for the transcription of genes as needed for survival.

Things to Watch Out For

Avoid making the question more complicated than it is. While we covered additional details about the *lac* operon in this explanation, the key point is that operons have multiple parts that each serve a specific function. Operons are complex, but can be broken down into simpler pieces.

E Membrane Transport

Sodium-potassium ATPase can be used by some somatic cells to control their volume as osmolarity within the circulatory system varies. A physician treats two patients that make poor dietary choices. Patient A suffers from hypertension (high blood pressure) that is exacerbated by a high-sodium diet. Patient B suffers from diabetes mellitus, which is poorly controlled due to a diet high in carbohydrates. How will the activity of sodium-potassium ATPase in these two individuals compare to the activity of this enzyme in a healthy individual?

Key Concepts

Biochemistry Chapter 8
Tonicity
Concentration gradients
Sodium–potassium pump
Osmosis

	Formula	High Solute Environment	Low Solute Environment	Equal Solute Environment	Applications
Molarity:	$\dfrac{Mol}{Liter}$	hypertonic	hypotonic	isotonic	Primarily in general chemistry
Osmolarity:	$\dfrac{Osmol}{Liter}$	hyperosmotic	hypo-osmotic	isosmotic	Primarily in biology

1 What type of environment exists within patients A and B when compared to a healthy individual?

Most cells in the body allow for free movement of water across cell membranes. Osmolarity, or concentration of all solute per liter of solution, is a measure of how dilute a fluid is. If the osmolarity around a cell does not match the osmolarity within the cell, water will move via osmosis to try to reestablish equilibrium. Patient A has a high-sodium diet and patient B has a high-carbohydrate diet, meaning the bodies of both patients must have a higher extracellular concentration of solute than is normal. Thus, the extracellular fluid of both patients would be hyperosmotic as compared to within a normal cell. You would expect the cells of both patients A and B to shrink as water leaves in an attempt to lower the osmolarity of the extracellular fluid.

Takeaways

Questions that involve osmosis are usually combined with other biology topics (particularly renal physiology), general chemistry topics (colligative properties), or biochemistry topics (membrane transporters) to necessitate multistep solutions. The key is to have a solid understanding of what the terms hypertonic and hypotonic mean and to be able to apply them to situations correctly.

2 Why does the activity of the enzyme decrease in patients A and B?

The sodium-potassium ATPase pump moves 3 Na^+ ions out of the cell for every 2 K^+ ions moved into the cell, at the cost of 1 ATP. The imbalanced ion movement functionally lowers the osmolarity of the cell in comparison to the environment, by moving more salt out than is taken in. In Patients A and B, the cell is already too dilute, or hypo-osmotic relative to the extracellular fluid. Moving additional salt out of the

K

cell would only serve to worsen the situation of Patients A and B. As a result, both patients would be expected to decrease function of the sodium–potassium pump.

Related Questions

1. Antidiuretic hormone (ADH) directly increases the ability of the blood to reabsorb water from the nephron. If an individual's blood becomes hypotonic with respect to the filtrate, would ADH secretion increase or decrease?

2. The reabsorption of water from the filtrate increases as the concentration of the interstitial fluid increases. Using the terms *hypertonic* and *hypo-osmotic*, describe the relationship between the interstitial fluid and the filtrate as well as the relationship between the filtrate and the interstitial fluid.

3. Alcohol and caffeine block the activity of ADH, a hormone that increases the ability of the blood to reabsorb water from the filtrate. An individual drinks a large cup of coffee in the morning and, when he goes to the restroom, finds that his urine is nearly colorless. Was the urine produced hypotonic, isotonic, or hypertonic to the blood?

Things to Watch Out For

Some students presume that the scenario presented will eventually return to equilibrium and may predict that ATP consumption will decrease and then increase again. However, the question stem does not guarantee a return to equilibrium. Be wary of trying to read too much into the question.

⊟ Glycolysis

Pyruvate kinase deficiency is a partial enzyme defect that often results in anemia caused by a decrease in the integrity of red blood cells. What is the purpose of this enzyme and why does a deficiency in its activity cause anemia?

1 What are the characteristics of the reaction catalyzed by pyruvate kinase?

Pyruvate kinase (PK) catalyzes the final reaction of glycolysis. While the focus in this question is on PK, remember that the other high-yield glycolytic enzymes to remember for Test Day are *hexokinase/glucokinase*, *phosphofructokinase-1* and *-2*, *glyceraldehyde-3-phosphate dehydrogenase*, and *3-phosphoglycerate kinase*.

Pyruvate kinase is a notable enzyme for three reasons. First, it catalyzes the formation of pyruvate from phosphoenolpyruvate (PEP) during the final reaction of glycolysis. Second, this reaction is one of the substrate-level phosphorylation steps during glycolysis, generating ATP from ADP and an inorganic phosphate. Because glucose is split into two three-carbon molecules during glycolysis, the conversion of PEP to pyruvate occurs twice for every glucose molecule and results in 2 ATP. Finally, this reaction catalyzed by pyruvate kinase is one of the irreversible reactions of glycolysis. Thus, pyruvate kinase cannot be used to generate PEP from pyruvate—this conversion requires *pyruvate carboxylase* and *phosphoenolpyruvate carboxykinase* (PEPCK) during gluconeogenesis.

2 What would be the effect of an inability to form pyruvate?

If an individual has a partial enzyme defect of pyruvate kinase, then significantly less pyruvate will be generated during glycolysis. Pyruvate kinase deficiency is the only common partial enzyme defect of glycolysis and is caused by a mutation in the *PKLR* gene. A variety of mutations in this gene can all cause reduced function of the pyruvate kinase enzyme. The *PKLR* gene produces an isoform of pyruvate kinase in various body tissues—most notably, the liver and red blood cells. Logically, these cells with reduced pyruvate kinase activity produce less pyruvate. In addition, because one of the substrate-level phosphorylation steps is lost, the net production of ATP per molecule of glucose drops from two to zero. In most cell types, this would mean that glucose could not be processed, but there would be little impact on the β-oxidation of fatty acids and other energy-producing pathways.

Why would the lack of pyruvate kinase be detrimental to red blood cells?

The question tells us that a person with pyruvate kinase deficiency often has anemia due to a decrease in the integrity of red blood cells. Why might this be? Mature erythrocytes lack mitochondria and rely solely on glycolysis for ATP production. In the absence of pyruvate kinase, red blood cells have little energy available for necessary processes such as membrane maintenance, leading to changes in cell shape. Ultimately, these misshapen cells are phagocytized by macrophages in the spleen. This premature destruction of red blood cells results in a shortage of red blood cells, or anemia. This anemia is exacerbated by the fact that some red blood cells may lyse within the vasculature because they cannot use the sodium–potassium pump to maintain cell volume effectively. Chronic hemolytic anemia also results in splenomegaly, or an enlarged spleen, excess iron in the bloodstream, and reduced oxygen transport.

Takeaways

Pyruvate kinase is an essential enzyme for red blood cells in particular. Biochemical pathways do not occur in isolation; each aberration in a pathway may have far-reaching consequences for the organism as a whole.

Things to Watch Out For

Many students, upon seeing this question, will recognize that the red blood cell will have difficulty making pyruvate; however, it is important to go one step further to determine the real reason why this is uniquely such a problem for the red blood cell.

Related Questions

1. Which step(s) in glycolysis require an *input* of ATP?
2. Which step(s) in glycolysis *produce* ATP?
3. Which step(s) in glycolysis catalyze the reduction of NAD^+?

Ⓢ Aerobic Metabolism

2,4-Dinitrophenol, also known as DNP, is a molecule that was once marketed as a weight loss "miracle" drug. DNP works by "uncoupling" oxidative phosphorylation from the electron transport chain, causing rapid weight loss. However, DNP also causes unpleasant, sometimes even fatal side effects, including tachypnea (elevated breathing rate) and hyperthermia. DNP was quickly pulled from the market as doctors realized that a fatal dose of DNP might be as little as twice the effective dose for weight loss. How does DNP cause weight loss and why might this mechanism explain the potentially fatal side effects of this drug?

❶ Why is DNP mostly deprotonated at physiological pH?

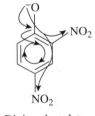

2,4-Dinitrophenolate
$pK_a = 4$

The hydroxyl group in DNP is substantially more acidic than a typical hydroxyl group. This is in large part due to that fact that the two nitro ($-NO_2$) substituents are very electron withdrawing. This withdrawing effect, facilitated by the inherently resonant structure of the aromatic ring, increases the acidity of the hydroxyl proton.

❷ Why can DNP cross physiological membranes despite being charged at physiological pH?

The two strongly withdrawing nitro groups, in combination with the benzene ring, help distribute the negative charge evenly across the DNP molecule. This charge dispersion ensures that negative charge is not concentrated at any one point and more-or-less balanced across the molecule, meaning that the usual barrier to charge (relative hydrophobicity) does not apply to DNP, even deprotonated (and therefore negatively charged) DNP.

3 Why is "coupling" of the electron transport chain and oxidative phosphorylation important for ATP production?

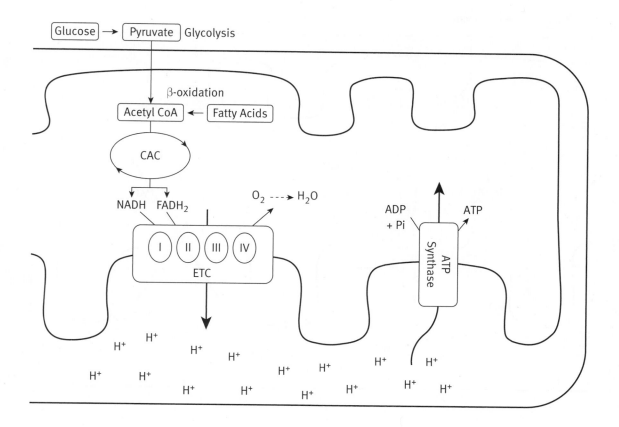

Glycolysis, pyruvate decarboxylation, and the TCA cycle produce several reducing agents capable of donating electrons into the electron transport chain. The series of reductions within the ETC ends with the final electron acceptor, oxygen. Over this series of reductions down the ETC, several protons are moved across the inner mitochondrial membrane from the matrix into the intermembrane space. This buildup of protons creates a concentration and charge gradient across the inner mitochondrial membrane. These protons then flow down their concentration gradient, through the specialized enzyme known as ATP synthase, which produces ATP. This ATP production is dependent on the presence of the electron transport chain, as ATP production relies on a differential concentration of protons to occur.

4 Why does protonating DNP within the intermembrane space disrupt chemiosmosis?

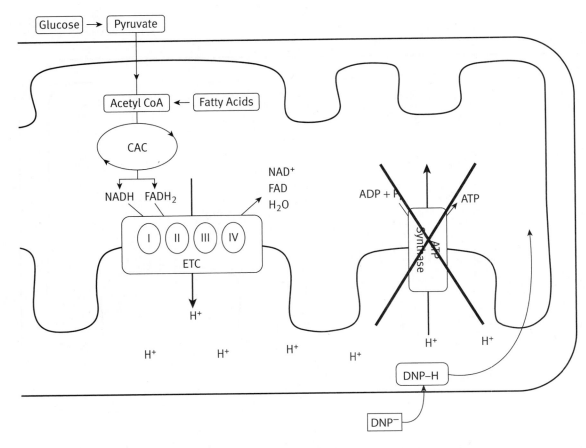

Takeaways

It is not sufficient to memorize the electron transport chain. Fundamental understanding of the electron transport chain and how a toxin may change its function is necessary to answer questions like these on Test Day.

DNP enters the intermembrane space and becomes protonated in the intermembrane space as a result of the excess of protons within the intermembrane space. As such, DNP sequesters protons thereby preventing these protons from powering ATP synthesis. In other words, DNP is removing the proton motive force by disrupting the proton gradient across the inner mitochondrial membrane.

Things to Watch Out For

The electron transport chain is nothing more than a series of oxidation–reduction reactions in which each complex has a higher reduction potential than the one that preceded it. This allows for a flow of electrons down the chain. Remember that oxidation is a loss of electrons and reduction is a gain.

5 Why does disrupting chemiosmosis cause weight loss and potentially fatal side effects?

If the ETC is running but not supplying sufficient protons to create a gradient (given that DNP is sequestering these protons), the ETC will continue to run at maximum rate in a futile attempt to continue to create a proton gradient. The energy that is being lost within the system as DNP moves protons without running the synthase pump will be lost as heat. A patient taking DNP would experience rapidly increasing body temperature due to the energetic favorability of moving protons down the gradient, as well as excessive use of energetic molecules as the mitochondria try to compensate for the effects of DNP.

Related Questions

1. Cyanide is a deadly protein that binds irreversibly to cytochrome ala3. What are the most likely immediate consequences of this disturbance?

2. If a toxin were administered that inhibited the electron transport chain at complex I, would the amount of ATP produced be more or less than if cyanide were administered?

3. What is one possible explanation why $FADH_2$ produces less ATP than NADH?

Key Concepts

Biochemistry Chapter 11

β-Oxidation

Electron carriers

ATP

Fatty acids

S Lipid Metabolism

Thermogenin is a protein that newborn humans use to generate heat through the oxidation of fatty acids by uncoupling the electron transport chain and oxidative phosphorylation within specialized fat cells. Because thermogenin activity leads to fatty acid metabolism, it also causes weight loss. Recently, researchers discovered that *adult* humans also express small amounts of thermogenin, and that lean adults tend to express more thermogenin, and obese adults less. One possible conclusion from this observation is that inducing thermogenin expression may be a treatment for obesity.

One group of researchers hypothesizes that thermogenin expression in adults is blocked by a suppressor protein. This group has filed a patent protecting their as-yet-unfinished method of inducing thermogenin expression using exogenous microRNAs, which are small segments of RNA complementary to a target gene. However, even if this group can successfully induce thermogenin expression using microRNA, they must also determine a method to make these microRNA molecules target only adipose tissue. Explain this group's plan to use microRNA to induce thermogenin expression. Why must thermogenin expression be limited to adipose tissue?

1 **By what mechanism would microRNA prevent the suppressor protein from suppressing thermogenin?**

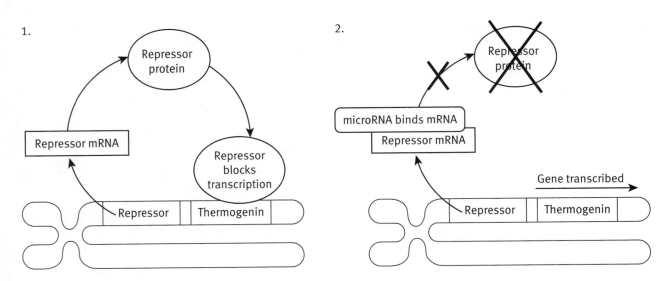

The suppressor protein must be coded for in some region of the DNA, most likely near the region that codes for thermogenin, as the suppressor protein is a regulatory protein for thermogenin. The microRNA cannot bind to this DNA region because DNA is double-stranded. However, the microRNA can bind to the mRNA transcript for the suppressor protein once that mRNA transcript has been transcribed. If the microRNA binds successfully to the mRNA transcript for the suppressor protein, that binding would prevent translation of the suppressor protein and lower the concentration of suppressor protein. Without the presence of the suppressor protein, thermogenin could now be expressed.

2 **Why would researchers seek to induce thermogenin expression only in adipose tissue?**

Decanoic Acid
MW = 172 g/mol

Carbons less oxidized,
more electrons for ETC.

Glucose
MW = 180 g/mol

Carbons already partially oxidized,
fewer electrons for ETC.

Adipose tissue stores and uses energy primarily in the form of fatty acids. Most other energetic tissue in the body preferentially utilizes glucose for energy generation. Compared to an equivalent molecular weight fatty acid (decanoic acid), glucose produces a far smaller quantity of acetyl CoA, and thus a far smaller quantity of energy. Decanoic acid's greater acetyl CoA production is due to the relatively high carbon density of decanoic acid, as compared to the relatively low carbon density (and high oxygen density) of glucose. The high quantity of oxygen in glucose indicates that glucose is already partially oxidized and has less reduction potential to contribute to the ETC. Production of thermogenin in adipose tissues would result in successful heat generation and use of this high-energy storage molecule. Comparatively, tissues burning glucose would only be able to produce a small quantity of heat from the uncoupling of the electron transport chain. Further, burning up glucose in this inefficient manner may cause normal tissues to become energy starved at a rate that is too high for healthy function of the body.

Takeaways

The energy content of fatty acids is much higher than that of sugars, due to the relatively high capacity for fatty acids to produce Acetyl-CoA and feed the electron transport chain.

Things to Watch Out For

Molecular weight alone is not a predictor of energy-generating capability in the mitochondrion. A better predictor would be number of carbons available for the creation of acetyl CoA.

Related Questions

1. β-Oxidation of a fatty acid yields nine molecules of NADH and nine molecules of $FADH_2$. How many carbons were initially present in the fatty acid, assuming it was saturated?

2. How many molecules of ATP would be generated by the complete oxidation of one molecule of palmitate (16:0)?

3. How many molecules of ATP would be generated by the complete oxidation of one molecule of arachidic acid (20:0)?

High-Yield Problem-Solving Guide questions continue on the next page. ▶ ▶ ▶

Key Concepts

Biochemistry Chapter 12

Postprandial (absorptive or well-fed) state
Postabsorptive (fasting) state
Insulin
Glucagon

$§$ Hormonal Regulation of Metabolism

Bodybuilders use a variety of anabolic substances to gain mass. Two such compounds are insulin and trenbolone. Insulin has significant anabolic and anti-catabolic properties and impacts the metabolism of various macromolecules, not just that of carbohydrates. Trenbolone binds the androgen receptor with an affinity five times higher than that of testosterone and is popular for its fat-burning and anabolic properties.

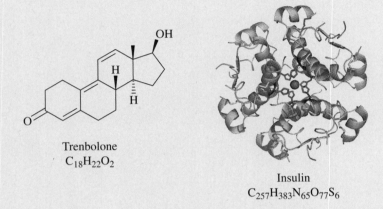

Trenbolone
$C_{18}H_{22}O_2$

Insulin
$C_{257}H_{383}N_{65}O_{77}S_6$

How do the different targets, mechanisms of action, and durations each ultimately lead to the same desired effect (increase in lean body mass)?

1 **Why is trenbolone able to permeate the cell membrane?**

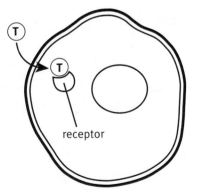

Trenbolone is a steroid hormone. Steroid hormones are derived from cholesterol, a cell membrane component. Trenbolone's net nonpolar structure allows it to travel through the cell membrane via passive diffusion. Trenbolone's receptors would thus be expected to be located within the cell.

K

2 **Once metabolized, trenbolone increases ammonium ion uptake in the muscle tissue. How does this promote mass gain?**

Following administration of trenbolone, ammonium uptake by muscle cells is increased. Ammonia (the deprotonated form of ammonium) is a precursor to protein formation. Increased quantities of ammonia available to the muscle allow for increased generation of protein. As protein synthesis in the muscles is upregulated, mass of the muscles is increased.

3 **Given insulin's structure, where is its target likely located?**

Insulin is a peptide chain hormone. It possesses hydrophilic and hydrophobic regions, and is also substantially larger than the phospholipids that comprise the membrane bilayer. Insulin is incapable of passively diffusing through the bilayer due to its size and hydrophilicity. Its receptors must be located in the cell membrane.

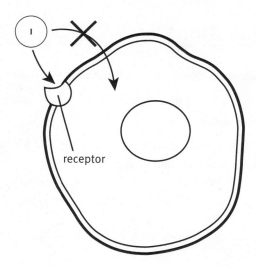

receptor

Takeaways

Hormonal regulation of metabolism is a commonly tested topic. Insulin, glucagon, and blood glucose levels all play a large role in regulation of various metabolic pathways.

Things to Watch Out For

While the MCAT can certainly test particular details of a given pathway, it behooves you to understand the main purpose of a given pathway to predict its regulation. Given that insulin is released when blood glucose concentrations are high, it only makes sense that insulin would be an activator of glycogen production (glycogenesis) and an inhibitor of glycogen breakdown (glycogenolysis).

4 **What can the negative symptoms of insulin be attributed to?**

Administration of insulin results in headache, nausea, hunger, confusion, and weakness. Insulin decreases blood availability of glucose via a number of pathways. If the individual is not in a postprandial state, the glucose suppression by insulin directly affects cells that would be using glucose during normal (postabsorptive) function.

With diminished quantities of glucose available to starved cells, side effects are hunger signaling and difficulty in function, as would be expected in a fasting state.

5 What metabolic processes does insulin favor for each energy source?

Insulin has major effects on muscle and adipose tissue. It increases the rate of glucose transport across the cell membrane, decreases the rate of lypolysis, and increases uptake of triglycerides and some amino acids from the blood. Essentially, insulin functions to ensure that, during times of high glucose availability, the body preferentially uses glucose for energy. Further, insulin signals for increased glucose storage and decreased use of alternative energy resources during the postprandial state. Expected processes would be as follows:

Carbohydrates: ↑ glycogenesis, ↓ glycogenolysis, ↓ gluconeogenesis (liver)

Lipids: ↓ lipolysis (adipocytes)

Proteins: ↑ protein synthesis (muscle tissue)

Related Questions

1. Identify the hormonal changes that occur during the postabsorptive state.

2. An insulinoma is a tumor that secretes high levels of insulin. What would be likely symptoms of this condition?

3. A glucagonoma is a tumor that secretes high levels of glucagon. What would be likely symptoms of this condition?

Solutions to Related Questions

1: Elements of Peptide Structure

1. Cysteine residues contain sulfhydryl groups that form disulfide bonds, a type of tertiary structure. Disulfide bonds can also stabilize local secondary and other tertiary structures as well. In addition, disulfide bonds help form a hydrophobic core within proteins, facilitating hydrophobic interactions. A change in a single cysteine residue may interrupt a key disulfide bridge, resulting in the inability of a protein to maintain its secondary and tertiary structures or its hydrophobic core. Ultimately, this may also disrupt quaternary structure by changing interactions between multiple peptides. Thus, this single change can cause disruption of protein structure and function.

2. 6 M HCl is concentrated solution of a strong acid. In fact, the pH of this solution is approximately –0.78. For comparison, gastric acid has a pH of 2 and a concentration of approximately 0.01 M HCl. Treatment of a protein with a strong acid will result in denaturation of the protein, or loss of elements of secondary, tertiary, and quaternary structure. Concentrated strong acid may also permit hydrolysis of the peptide bond, interrupting primary structure.

3. This question combines concepts from both amino acid structure and peptide function. A G protein-coupled receptor contains a transmembrane domain that consists mainly of hydrophobic residues. Because this mutation is not in the binding site of the receptor, it is unlikely to affect substrate binding. The specific mutation described substitutes one hydrophobic amino acid for another, which should cause minimal changes in the secondary and tertiary structures of the protein. Therefore, this mutation is unlikely to affect the structure or function of the peptide.

2: Enzyme Kinetics

1. A noncompetitive inhibitor binds to an allosteric site, rather than the active site. Binding of the inhibitor to an allosteric site changes the conformation of the active site, leading to a decrease in the efficiency of enzyme catalysis, which results in a decrease in V_{max}. However, K_m will not change due to noncompetitive inhibition because any copies of the enzyme still in the active conformation can bind the substrate with the same affinity. Therefore, compared to the line without inhibitor, the line with a noncompetitive inhibitor will have the same x-intercept, but a higher y-intercept. This is shown in the diagram below:

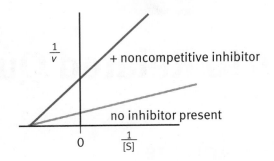

2. An uncompetitive inhibitor binds to an allosteric site, but only when the substrate is already bound to the enzyme. This results in an increased affinity for substrate bound to enzyme and decreased dissociation of the enzyme–substrate complex. Therefore, uncompetitive inhibition results in a decrease in both V_{max} and K_m. Compared to the line without an inhibitor, the line with an uncompetitive inhibitor will have a more negative x-intercept and higher y-intercept—a shift upwards and to the left, which is shown in the diagram below:

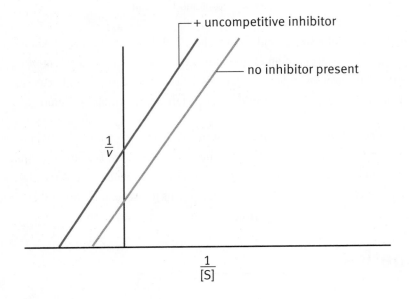

3. Michaelis–Menten plots contain the same information as Lineweaver–Burk plots, but have different axes. The y-axis in a Michaelis–Menten plot is reaction rate or velocity, V, and the x-axis is substrate concentration, [S]. Monomeric enzymes will result in a hyperbolic shape in the graph, reaching a plateau at V_{max}. A competitive inhibitor will not affect V_{max}; however, the curve will be stretched to the right because K_m increases, as shown in the diagram below:

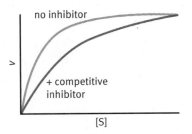

3: Amino Acid Electrophoresis

1. If a polypeptide is in an environment with a lower pH than its pI, then this implies that it is in a relatively acidic environment compared to its pI and the polypeptide will be more protonated than at its pI. This means that it will take on a positive charge. Cations always migrate toward the cathode in any type of electrochemical cell.

2. In a gel with a constant pH, amino acids will exist either in their positive, zwitterion, or negative forms. If the isoelectric point of a given amino acid is equal to the gel's pH, it will remain stationary. Otherwise, it will move either to the anode or cathode with a constant speed. Conversely, on a gel with a pH gradient, as the amino acid travels toward the anode or the cathode, the pH will change. Accordingly, the amino acid will either protonate or deprotonate until it reaches its zwitterion form, at which point it stops moving on the gel.

4: Isomerism in Carbohydrates

1. Anomers are cyclic epimers that differ in chirality at the anomeric carbon. Mutarotation occurs about the anomeric carbon, which arises when a hydroxyl group in the sugar attacks the carbonyl carbon, leading to ring closure through a nucleophilic addition reaction. The position of the –H and –OH groups on C-1 relative to the free –CH$_2$OH group in the sugar dictates whether the sugar is α or β. The molecules shown below are the two anomers of D-glucose.

α-D-glucopyranose β-D-glucopyranose

2. Cellulose, or dietary fiber, cannot be digested by humans and acts as a bulking agent in feces. Monosaccharides can traverse the walls of the intestine to enter the cells of the epithelium that lines the digestive tract. Polysaccharides must be enzymatically broken down into monosaccharides to be absorbed by the digestive tract. Cellulose is a large polysaccharide and humans lack the enzymes required to break the β-1,4 bonds that hold the compound together. This is reflective of enzyme specificity of the active site: human enzymes can break down α-1,4 bonds, but not β-1,4 bonds.

3. Epimers are sugar diastereomers that differ in configuration at exactly one chiral carbon. The first structure, fructose, contains a carbonyl group at C-2. None of the other structures contain a carbonyl on C-2, so fructose has no epimers in this group. Next, compare D-glucose and D-galactose. These two sugars differ only at C-4; therefore, D-glucose and D-galactose are C-4 epimers. Next, a comparison of D-glucose and D-mannose reveals inverted configuration at C-2, making them C-2 epimers. Finally, comparison of D-galactose and D-mannose reveals inverted configuration at C-2 and C-4, which means that these molecules are diastereomers of each other, but not epimers.

5: Lipid Saponification

1. Soap is a carboxylate salt of a fatty acid ionically bonded to a cation. Soap consists of two parts: a hydrophilic, negatively charged carboxylate group with a high affinity for water, and a nonpolar, hydrophobic alkyl tail that interacts with organic materials, such as soil and grease. The buildup is likely composed of fatty acids that are bonded to other cations that reduce its solubility in water, such as calcium (like the example with pancreatitis) or magnesium. In addition to hard water decreasing the solubility of soap, water that is acidic can reduce the effectiveness of soap. Acidic solutions contain large quantities of protons and will rapidly protonate the negatively charged carboxylate groups of the fatty acids. This reduces the hydrophilicity of these groups, reducing the effectiveness and solubility of soap.

2. The question stem states that adipocere or grave wax occurs in a cold, humid, low-oxygen environment. Bodies quickly begin decomposing after death as bacteria begin to digest the corpse. A low-oxygen environment favors anaerobic bacterial species. These anaerobes carry out saponification of fatty tissues in the body, leading to the formation of a chalky substance in and around fatty tissues.

3. Treating a triacylglycerol molecule with an acid instead of a base could still result in hydrolysis of the ester bonds; however, the carboxylate ions would undergo rapid protonation to form carboxylic acids, resulting in diminished interactions between the polar heads and water. This would decrease the solubility of the resulting carboxylic acids relative to the carboxylate anions. These compounds could not be considered soap. Soap, by definition, is composed of carboxylate salts—not carboxylic acids.

6: Central Dogma of Genetics

1. Human DNA polymerase is not used for PCR because of the heating and cooling processes involved. PCR requires enough heat to be transferred to the DNA molecules to cause them to denature and separate into two single strands. This amount of thermal energy would also likely denature human DNA polymerase, rendering it unable to carry out replication. The genus name of *T. aquaticus*, *Thermus*, is reflective of the fact that this bacterium's DNA polymerase is active at very high temperatures and is able to carry out replication even under the harsh conditions required in PCR.

2. The lagging strand is significantly more likely to acquire mutations than the leading strand. The lagging strand is synthesized in a discontinuous manner as Okazaki fragments, which must subsequently be coupled together by DNA ligase. This means that the lagging strand contains many more RNA primers, which must be removed and substituted for by DNA. This stop-and-start synthesis pattern and the fact that the lagging strand requires more postreplicatory modifications means that the lagging strand is more prone to mutations.

3. While the ^{15}N-labeled guanine question given in this chapter showed each of the daughter strands individually, there is another way to approach this question. Each of the two original unlabeled strands will end up in only one of the resulting DNA molecules; thus, only two of the eight resulting DNA molecules will contain an unlabeled strand.

7: Operons

1. Glucose is the preferred fuel for most organisms. As such, even in the presence of lactose, there is no need to process this disaccharide if glucose is plentiful. When there is glucose present, adenylate cyclase is inactive, which means that cAMP is not available to bind to CAP. Without the cAMP–CAP complex, RNA polymerase cannot bind strongly to the DNA. By extension, transcription of the genes for lactose digestion will not occur.

2. Inducible genes require the presence of a compound known as an inducer. In other words, the genes of the operon are only transcribed in the presence of a particular substrate—in the example in the question, this substrate is lactose. Repressible systems are repressed in the presence of a specific substrate and allow for certain genes to be turned off under particular conditions. Repressible systems are examples of negative feedback, in which the presence of the product of a pathway suppresses the pathway itself. This mechanism helps prevent wasteful energy use by the cell.

3. The *trp* operon is a repressible system. In the absence of the amino acid tryptophan, the repressor is unable to bind to the regulator sequence. Because there is no repressor, the genes required for tryptophan synthesis are available for transcription. In repressible systems, the default setting for the gene is *on*. When tryptophan is present, it binds to the repressor and the tryptophan–repressor complex binds to the operator region, halting transcription of the genes required for tryptophan synthesis.

8: Membrane Transport

1. If an individual's blood becomes hypotonic with respect to the filtrate, then the body would aim to increase free water excretion to regain homeostasis. This means that the kidney would decrease the amount of water reabsorbed from the filtrate into the blood. This could be accomplished by decreasing the secretion of ADH, which would promote the loss of more water in the urine, increasing the blood concentration.

2. If the concentration of the interstitial fluid is higher than that of the filtrate, then the interstitial fluid is hypertonic (or hyperosmotic) relative to the filtrate. The filtrate, then, is hypotonic (or hypo-osmotic) to the interstitial fluid.

3. If the activity of ADH is blocked, then it is unable to carry out its function—concentrating the urine by promoting the reabsorption of water from the filtrate into the interstitial fluid. Thus, after drinking coffee, an individual would excrete larger volumes of water. The fact that the urine is clear is indicative of its low concentration. When water cannot be reabsorbed from the collecting duct, the urine will be less concentrated (hypotonic) relative to the blood.

9: Glycolysis

1. ATP is consumed during two steps in glycolysis. The first is the conversion of glucose to glucose 6-phosphate by hexokinase or glucokinase. The second ATP-requiring step occurs during the conversion of fructose 6-phosphate to fructose 1,6-bisphosphate by phosphofructokinase-1 (PFK-1). Both of these steps involve the phosphorylation of sugars; ATP provides the phosphate required for these reactions.

2. ATP is produced through substrate-level phosphorylation in two different reactions. The first occurs during the conversion of 1,3-bisphophoglycerate to 3-phosphoglycerate by phosphoglycerate kinase. The second is the conversion of phospho-enolpyruvate to pyruvate by pyruvate kinase. Notice that each of these steps involves the removal of a phosphate from an intermediate of glycolysis. It is important to remember that 2 ATP are produced in each of these reactions for every glucose that enters glycolysis. This is because fructose 1,6-bisphosphate is cleaved into dihydroxyacetone phosphate and glyceraldehyde 3-phosphate.

3. The conversion of glyceraldehyde 3-phosphate to 1,3-bisphosphoglycerate by glyceraldehyde-3-phosphate dehydrogenase results in the reduction of NAD^+ to NADH. Like the substrate-level phosphorylation reactions described previously, 2 NADH are generated for every glucose that enters glycolysis. NADH can then feed into the electron transport chain to be used in oxidative phosphorylation.

10: Aerobic Metabolism

1. Cytochromes are essential for the electron transport chain. Because cyanide binds to, and inhibits, one of these key cytochromes, cyanide will shut down the entire electron transport chain. Therefore, the electron transport chain will stop creating a proton gradient. Without the mitochondrial proton gradient, the mitochondrion will no longer be able to produce ATP. However, unlike the case of 2,4-dinitrophenol (DNP), metabolism of biomolecules will also be inhibited because cyanide shuts down the electron transport chain, which shuts down the citric acid cycle as well.

2. The amount of ATP produced if the electron transport chain were inhibited at complex I is, perhaps counterintuitively, greater than if the chain were inhibited at complex IV (such as by cyanide). Complex I is responsible for the oxidation of NADH and the pumping of some protons into the intermembrane space. While inhibition of complex I would stop this electron carrier from feeding electrons into the chain, $FADH_2$ could still provide electrons because it transfers them to Complex II, after the point of inhibition. This would still allow creation of a proton-motive force using Complexes III and IV, with no buildup of electrons at the end of the chain. Thus, ATP could still be produced from $FADH_2$, even if it cannot be made from NADH.

3. As described previously, $FADH_2$ donates electrons to the electron transport chain at Complex II, while NADH donates electrons at Complex I. Assuming no inhibition occurs, protons are pumped into the intermembrane space, thereby increasing the proton-motive force. The proton-motive force is directly proportional to the energy stored in the concentration gradient; therefore, the larger the proton-motive force is, the more energy available for generating ATP.

11: Lipid Metabolism

1. In order to yield nine molecules of NADH and nine molecules of $FADH_2$, there must be nine rounds of oxidation. Using the same equation given in this question to determine the number of cycles (or mathematical deduction), the number of carbons can be found:

$$\frac{n}{2} - 1 = 9$$
$$n = 20$$

Thus, the original fatty acid must contain 20 carbons. This is arachidic acid (20:0).

2. β-Oxidation of palmitate would yield 7 NADH and 7 FADH$_2$, as described in the question. Each NADH can generate 2.5 ATP, whereas each FADH$_2$ can result in 1.5 ATP. Thus, there are $7 \times 2.5 + 7 \times 1.5 = 28$ ATP generated from the molecules of NADH and FADH$_2$. In addition, each molecule of acetyl-CoA can generate 3 NADH, 1 FADH$_2$, and 1 molecule of GTP (which is easily converted to ATP) during the citric acid cycle, resulting in an additional $3 \times 2.5 + 1 \times 1.5 + 1 = 10$ ATP per acetyl-CoA. β-Oxidation of palmitate yields 8 acetyl-CoA molecules, which results in 80 molecules of ATP. Therefore, in total, $28 + 80 = 108$ molecules of ATP will be created by complete oxidation of one molecule of palmitate. However, one molecule of ATP was required at the beginning to activate palmitate to palmitoyl-CoA. Therefore, the final yield is $108 - 1 = 107$ ATP.

3. First, the number of oxidation cycles, acetyl-CoA molecules, and NADH and FADH$_2$ molecules produced must be determined. Using the same logic as in the previous questions, β-oxidation of arachidic acid will result in nine NADH (22.5 ATP) and nine FADH$_2$ (13.5 ATP), for a total of 36 ATP. Because there are 20 carbons in arachidic acid, 10 acetyl-CoA molecules will result. Each acetyl-CoA molecule will yield 10 ATP, resulting in 100 ATP. The total quantity of ATP produced from complete oxidation of arachidic acid is 136 ATP. However, one molecule of ATP as required for activation of arachidic acid, making the total yield 135 ATP.

12: Hormonal Regulation of Metabolism

1. The postabsorptive (fasting) state occurs when blood glucose concentrations start to drop following a meal. Glucagon and epinephrine levels rise. In the liver, glycogen is broken down and the resulting glucose is released into the bloodstream. Gluconeogensis is also upregulated in response to glucagon, but this response is much slower. Amino acids are released from skeletal muscle, and free fatty acids are released from adipose tissue in response to increased epinephrine and decreased blood glucose. Amino acids and free fatty acids are taken up by the liver, where the amino acids provide the carbon skeletons and the oxidation of fatty acids provides the ATP required for gluconeogenesis. Concomitantly, pathways such as glycogenesis and lipid synthesis are inhibited.

2. An insulinoma releases excess insulin. This excess release of insulin results in low blood glucose levels (hypoglycemia). Because the brain cannot function properly at low blood glucose concentrations, altered mental status may occur. In addition, low blood glucose levels affect the nervous system, which can result in headache, fatigue, double vision, and blurring of vision. It is also important to note that as blood glucose levels fall, epinephrine is released, which can lead to tremors, palpitations, sweating, hunger, anxiety, and nausea.

3. A glucagonoma releases excess glucagon, causing elevated blood glucose levels (hyperglycemia). Normally, insulin has a suppressive effect on the secretion of glucagon. The marked increase in glucagon will cause breakdown of triacylglycerols in adipose tissue, resulting in the release of glycerol and free fatty acids into the bloodstream, which can ultimately result in weight loss. Patients with this condition often present with diabetes mellitus because of the body's inability to manage blood glucose levels, despite a compensatory increase in insulin secretion.

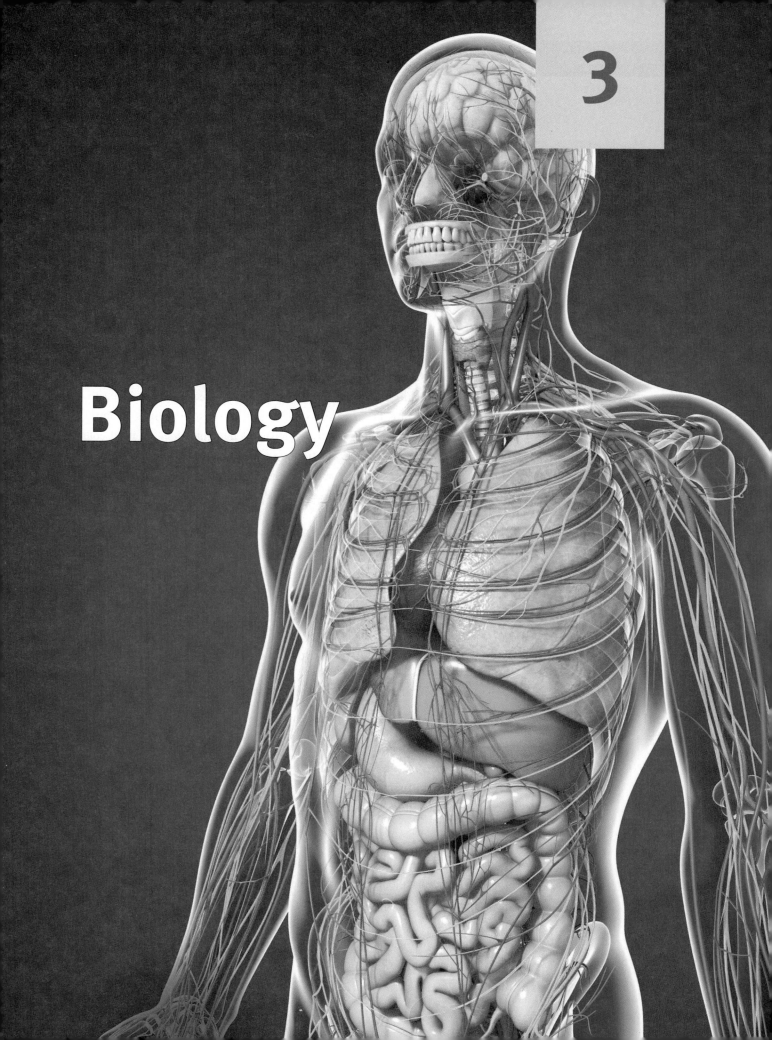

Biology

3

🄴 Prokaryotic Genetics

Bacterial genetic recombination can result in enhanced resistance to antibiotics. What bacterial genetic recombination process(es) is/are most likely to quickly convert entire bacterial colonies from being sensitive to a given antibiotic to being resistant to that antibiotic?

❶ What Plan should be used to answer this question?

This question is asking for identification of various methods used by bacteria for genetic recombination, and which of these processes may be most suited to the widespread acquisition of genes that confer antibiotic resistance in a colony. This indicates that this question requires two levels of thought. First, we need content knowledge regarding bacterial genetic recombination. Then, we must think critically about these processes with an eye toward recognizing what processes are most favorable for the rapid and widespread acquisition of antibiotic resistance.

Note that while there are two prokaryotic kingdoms—Archaea and Bacteria—our focus in this question is on bacteria, specifically.

❷ What are the bacterial genetic recombination processes?

Overall, there are three main methods of exchanging genetic information between prokaryotic cells: transformation, conjugation, and transduction.

Transformation occurs when a prokaryote picks up a piece of foreign DNA from the environment and integrates this DNA into its genome. Once integrated, this foreign DNA will be replicated in all subsequent rounds of binary fission and will thus be passed on to all daughter cells of this prokaryote. Transformation is a particularly common method of genetic recombination in gram-negative bacilli.

Conjugation is a prokaryotic form of sexual reproduction in which a conjugation bridge is formed between two cells. One cell, called a donor male (+), forms a sex pilus that can attach to a recipient female (–) cell. Once the sex pilus has fused with the recipient cell, genetic material can be passed through the conjugation bridge. This transfer is unidirectional, from the donor male to the recipient female. To form the sex pilus, a cell requires a sex factor, the best studied of which is the fertility (F) factor. During conjugation, a plasmid may be passed through the conjugation bridge; the

F factor is commonly transferred, turning the recipient female into a donor male cell as well. Other times, the donor male can attempt to donate its entire genome to the recipient female.

Finally, transduction is the transfer of genetic material from one prokaryote to another using a bacteriophage as an intermediate. This is an accidental method of genetic recombination because it arises from improper packaging of genetic material during the formation of new virions within a bacterium. First, a bacteriophage infects a bacterium, and its genetic material is replicated as new virions are assembled within the cell. As these viral genomes are packaged into capsids, small segments of bacterial DNA may accidentally be trapped within the capsid. The bacterium ultimately lyses, releasing these virions. When the virion carrying bacterial genetic information infects another bacterium, the transfer is completed; the second bacterium receives genetic material from the first via the bacteriophage vector.

 ### Which process is most favorable for transferring antibiotic resistance?

Which of these methods can transfer genetic material for antibiotic resistance? In short: all of them. Each process can transfer genetic material that codes for antibiotic resistance mechanisms. However, two of these processes are fairly random: transformation and transduction. Transformation requires the bacterium to encounter the genetic material for antibiotic resistance in its environment. While it is entirely possible for this to occur for individual bacteria, it is unlikely to convert an entire established bacterial colony from being antibiotic sensitive to being antibiotic resistant. Similarly, transduction requires the accidental removal of bacterial genetic information followed by packaging into phages. While this may convert individual bacteria from being antibiotic sensitive to being antibiotic resistant, it is unlikely to convert an entire colony.

Conjugation, on the other hand, is a deliberate and specific transferal of genetic material from one bacterium to another. Further, in many cases of conjugation, the recipient cell acquires the necessary sex factor to carry out conjugation with other cells; that is, the recipient is converted into a donor. This would allow rapid, exponential spread of genetic material between bacterial organisms. If the sex factor were located on the same plasmid as an antibiotic resistance gene, then this would allow the rapid conversion of an entire preexisting colony from being antibiotic sensitive to being antibiotic resistant.

Takeaways

Antibiotic resistance may be acquired through any of the three major prokaryotic genetic recombination processes, but the rate and ubiquity of the acquisition varies between the three processes. Transformation and transduction both occur on the individual level, whereas conjugation can convert entire colonies.

Things to Watch Out For

The wording of this question is important—it implies that the bacterial colonies acquiring antibiotic resistance in this question already exist. Transformation and transduction could result in whole colonies with antibiotic resistance—but these colonies would have to be descended from an individual cell that underwent recombination.

Related Questions

1. A pharmaceutical company would like to induce the production of a target protein by a particular strain of *E. coli*. What would be the required characteristics of the plasmid in order to create large colonies of *E. coli* that produce the target protein?

2. A researcher discovers that a type of bacteriophage removes the exact same sequence of DNA from each bacterium infected. What does this information likely indicate about the bacteriophage's ability to feed into either the lytic or lysogenic cycle?

3. It is noted by a researcher that a certain colony of bacteria has developed an extremely high rate of recombination, acquiring multiple phenotypic changes in a very short period of time. What mechanism related to conjugation would most likely explain this finding?

High-Yield Problem-Solving Guide questions continue on the next page. ▶ ▶ ▶

S The Menstrual Cycle

Chronic anovulation, the consistent absence of ovulation in a reproductive-age woman, is one of the leading causes of female infertility. 70% of chronic anovulation cases are due to hormone imbalances, usually insufficient FSH or LH levels. In these cases, the drug clomiphene citrate, trade name Clomid, is used to successfully and reliably induce ovulation. Clomid works by blocking estrogen receptors within the hypothalamus and the pituitary. Clomid is taken monthly in a five-day course. The timing of the dosing course is important: If "day 1" of the menstrual cycle is the first day of menstruation, Clomid should be taken from days 5–9 of the cycle. If Clomid is to be used to successfully induce ovulation, why should Clomid be taken during this part of the menstrual cycle?

1 Describe the hormones in the HPA axis and their ultimate effect on the ovaries.

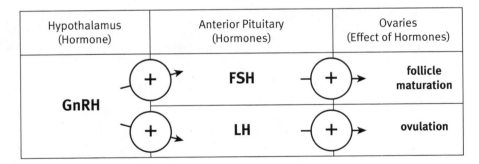

Hypothalamus (Hormone)	Anterior Pituitary (Hormones)	Ovaries (Effect of Hormones)
GnRH (+)	(+) FSH (+)	follicle maturation
(+)	(+) LH (+)	ovulation

During the follicular phase (starting around day 5), GnRH (Gonadotropin Releasing Hormone) is produced by the hypothalamus. This release of GnRH stimulates the release of FSH (Follicle Stimulating Hormone) from the anterior pituitary. FSH travels via the blood to the ovaries, where it stimulates maturation of the follicle and ovarian release of estrogen. This estrogen negatively feeds back on the GnRH release from the hypothalamus, preventing further release of FSH. As day 14 is approached, GnRH is again released, this time to stimulate the production of LH (Luteinizing Hormone) from the anterior pituitary. LH travels via the blood to the ovaries, where it stimulates the rupture of the ovarian follicle, creating the corpus luteum and facilitating the release of progesterone.

2 **Why must estrogen have an inhibitory effect on the GnRH hormone cascade during follicular maturation?**

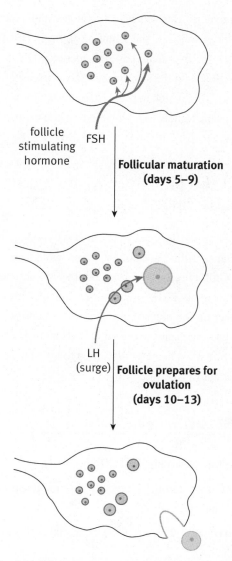

follicle stimulating hormone — FSH

Follicular maturation (days 5–9)

LH (surge) — **Follicle prepares for ovulation (days 10–13)**

Early in the follicular phase, estrogen acts on the uterus, causing vascularization of the endometrium. Estrogen also acts in a negative feedback loop to in order to inhibit further FSH release from the anterior pituitary. This FSH inhibition is necessary to prevent the development of multiple egg follicles during the maturation phase.

Takeaways

It is important to have a good understanding of the normal way that systems such as the menstrual cycle function. Using that knowledge, different variables, such as disease or dysfunction, can be applied to the system, and the results of that dysfunction can be found in a methodical way.

Things to Watch Out For

Estrogen has both negative and positive feedback effects on FSH and LH at different times in the menstrual cycle. Remember that estrogen levels fall dramatically after the LH surge but rise again during the luteal phase. During this phase, however, both estrogen and progesterone are now produced by the corpus luteum, and both have a negative feedback effect.

3 Why must estrogen's effect on the GnRH cascade switch to excitatory in the days leading up to ovulation?

As day 14 of the cycle approaches, estrogen levels begin to rapidly increase as estrogen is being released in greater quantity by the maturing follicle. This burst of estrogen surpasses a threshold within the hypothalamus and anterior pituitary, so that estrogen begins to exert positive feedback on the anterior pituitary, increasing the sensitivity of LH releasing cells to GnRH. This switch to positive feedback thus allows for a surge in LH levels, which stimulates the follicle to rupture and release the egg.

4 Why should the patient take Clomid from days 5–9 and discontinued after day 10?

Clomid works to block estrogen receptors within the hypothalamus and anterior pituitary. During follicular maturation, taking Clomid would prevent the negative feedback of estrogen on GnRH levels, which would thus increase levels of FSH and thus assist with low endogenous FSH levels in the female. However, during pre-ovulation, estrogen's effect on the anterior pituitary is excitatory, meaning that the administration of Clomid would decrease output of FSH and LH during days 10–13. Thus, Clomid administration must be carefully timed to avoid reducing the levels of the target hormones FSH and LH.

Related Questions

1. At what point in the follicular phase is FSH inhibited?

2. How do the levels of progesterone change during the menstrual cycle? What is the function of progesterone?

3. How could ovulation be prevented during the menstrual cycle through biochemical means?

High-Yield Problem-Solving Guide questions continue on the next page. ▶ ▶ ▶

Key Concepts

Biology Chapter 3
Neurulation
Gastrulation
Implantation
Cleavage

⊟ Stages of Embryogenesis

Hirschsprung's disease is a congenital disorder that occurs when neural crest cells fail to complete their migration to the digestive tract. That is, the neural crest cells that should migrate the farthest under normal circumstances never reach their final location. How might this affect function?

1 What are the functions of neural crest cells?

Neural crest cells arise from the tips of the neural folds. As the neural folds fuse in the midline, they form the neural tube; the neural crest cells are displaced from the tube and migrate to various sites in the body. These cells differentiate into the peripheral nervous system (including the sensory ganglia, autonomic ganglia, adrenal medulla, and Schwann cells), as well as specific cell types in other tissues (such as calcitonin-producing cells of the thyroid, melanocytes in the skin, and others).

2 What is the pattern of neural crest cell migration?

Neural crest cells migrate outward from the tip of each neural fold. They enter the walls of the digestive tract in a cranial-to-caudal fashion, meaning that they innervate the portions of the tract closest to the mouth first and the portions of the tract closest to the anus last.

3 How would a lack of neural crest cells affect digestive tract function?

In Hirschsprung's disease, the neural crest cells destined for the distal portions of the digestive tract never reach their location, and thus cannot form the parts of the enteric nervous system that would normally exist in these parts of the gut tube. Hirschsprung's disease can affect a variable length of the gut tube starting from the anus and including some length of the large intestine.

Because the neural crest cells ultimately form both sensory and autonomic ganglia, both of these modalities would be lost. This means that there would be reduced sympathetic and parasympathetic responses in the digestive tract (slowed or absent peristalsis, reduced secretion) as well as a loss of visceral sensation, making it difficult for a person with this diagnosis to sense when he or she needs to defecate.

Takeaways

Neural crest cells originate from the tips of the neural folds and migrate to form a diverse group of tissues, including the peripheral nervous system, calcitonin-producing cells of the thyroid, melanocytes, and others.

Things to Watch Out For

Neural crest cells migrate into the wall of the digestive system in a cranial-to-caudal (mouth-to-anus) fashion. Note that this is the opposite direction as the formation of the gut tube itself, which grows from anus to mouth in deuterostomes, such as humans.

Related Questions

1. Gestational trophoblastic neoplasms include choriocarcinoma, a highly invasive cancer that has often widely metastasized by the time of diagnosis. Why might choriocarcinoma have a propensity to invade and metastasize?

2. A mutation occurs in the mesoderm very early in gastrulation. This mutation is then propagated through many cell divisions. What organ systems are likely to be affected by such a mutation?

3. A zygote undergoes indeterminate cleavage to form two separate zygotes. What is a possible outcome of this division?

S Action Potentials

Hyperkalemia is a condition characterized by elevated concentration of the electrolyte potassium in the blood, and consequently also in the extracellular fluid. Because of potassium's pivotal role both in establishing the resting potential and in controlling the action potential, hyperkalemia may disrupt nervous system activity. Hyperkalemia can arise as a possible complication due to the maltreatment of an athlete suffering from an imbalance of electrolytes. An athlete may lose large quantities of potassium through diaphoresis (sweating) during long periods of exercise. This potassium may be replaced using an intravenous administration of a KCl saline solution. However, if the KCl solution is administered too rapidly, the rapid influx of potassium ions into the athlete's circulatory system may cause hyperkalemia instead. If an athlete does begin to suffer from mild hyperkalemia, how would the resting potential and action potential be affected?

1 How does potassium efflux help establish the resting potential in a neuron?

Potassium flows from the intracellular to the extracellular space via facilitated diffusion. Potassium leaking out of the cell causes the inside of the cell to become more negative because potassium carries a positive charge. In doing so, potassium follows its concentration gradient. Sodium's concentration is higher outside the cell, and therefore it seeks to follow its concentration gradient by flowing into the cell. However, there are substantially more leak channels for K^+ as compared to leak channels for other ions, including Na^+. Thus, while sodium leak channels do allow some sodium to leak in, which slightly increases the potential of the inside of the cell, conductance in resting cells is only high for potassium. This high K^+ conductance explains why the resting potential of the cell is very close to potassium's equilibrium potential (–85 mV for physiological intra- and extracellular concentrations). The Na^+/K^+ pump's role is to reestablish the concentration gradients of sodium and potassium by pumping the former out of the cell and the latter into the cell.

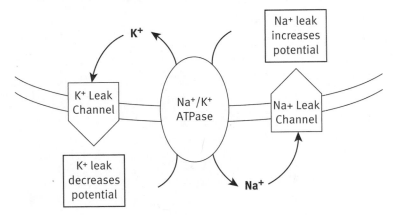

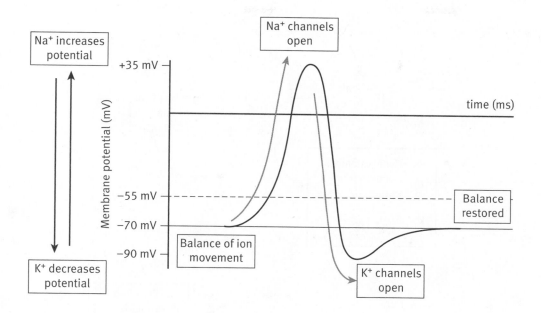

2 What role does potassium play in the action potential itself?

The resting potential of the neuronal cell is –70 mV. External stimuli cause sodium channels to open, thereby allowing sodium to flow into the cell. Once the membrane potential reaches –55 mV, all sodium channels open, leading to a rush of sodium ions into the cell. As soon as the membrane potential climbs to +35 mV, all sodium channels close and potassium channels open. The opening of K^+ channels leads to potassium efflux, lowering the cell's membrane potential all the way –85 mV. Therefore, potassium's role is repolarization (+35 mV to –70 mV) and then hyperpolarization (–70 mV to –85 mV).

3 Why would hyperkalemia raise the resting potential?

Hyperkalemia raises the resting potential because potassium is the biggest contributor to the resting potential. In normal circumstances, the concentration gradient of potassium leads to potassium efflux, which contributes to lowering the membrane potential. However, in hyperkalemia, the concentration of potassium ions outside the cell is heightened. Therefore, the normal concentration gradient that leads to potassium ion efflux is disrupted, and potassium ions no longer exit the cell. This phenomenon leads to a heightened resting membrane potential due to lack of potassium efflux.

Takeaways

While the potassium channels involved in the action potential can be open or closed, the sodium channels have three possibilities: open, closed, or inactivated. Inactivation occurs at positive potentials and the channel must be brought down below threshold to be deinactivated. Note that when the channel is deinactivated, it is still closed, not open; the membrane potential will have to be brought to threshold again for this channel to be opened.

Things to Watch Out For

Hyperkalemia is not a physiologically normal state. Remember that when intra- and extracellular ion concentrations are in their normal ranges, the shape of an action potential always remains consistent. Moreover, action potentials are considered all-or-nothing phenomena because once threshold is reached, an action potential will continue through the same stereotyped pattern of electrical potential changes, regardless of the identify or magnitude of the initiating stimulus.

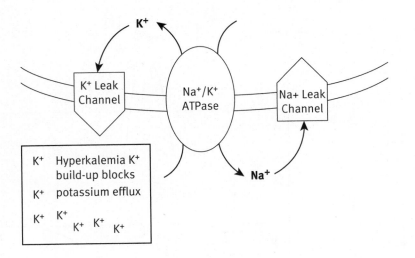

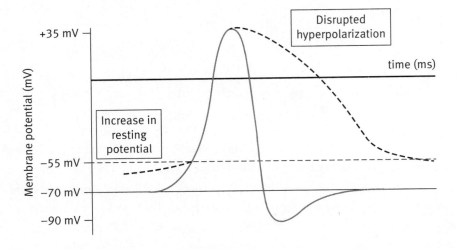

4 How would the period after depolarization be affected by hyperkalemia?

The high extracellular concentration of potassium ions outside the cells blocks the efflux of potassium that normally occurs during repolarization. Therefore, when the potassium channels open at +35 mV, the efflux of potassium ions is much slower, and repolarization takes longer to occur.

Related Questions

1. At what point during the action potential is sodium closest to its electrochemical equilibrium?

2. How does a neuron encode the magnitude of a stimulus?

3. How could an action potential be inhibited?

High-Yield Problem-Solving Guide questions continue on the next page. ▶ ▶ ▶

Key Concepts
Biology Chapter 5
Anterior pituitary
Hypothalamus
Thyroid
Negative feedback

⑤ The Endocrine System

Depression is one symptom of a disease called Cushing syndrome. But where most cases of depression are treated over months using cognitive behavior therapy, depression in Cushing patients can be resolved in days using cortisol receptor blockers. Other classic symptoms of Cushing syndrome include acne, weakened muscles, and fatty deposits in the face and neck (called "moon facies" and "buffalo lumps"). Finally, in addition to these classic symptoms, some patients, though not all, may experience a darkening of the skin called hyperpigmentation. Interestingly, all of the symptoms of Cushing syndrome are resolved by cortisol receptor blockers, except hyperpigmentation.

Cushing syndrome is often caused by a hormone-secreting adenoma, which is a noncancerous tumor (tissue overgrowth) that behaves just like the tissue that it has grown out of. Any one of the following three tumor types can cause Cushing syndrome: (1) A tumor in the anterior pituitary; (2) A tumor in the adrenal cortex; or (3) An "ectopic" ACTH-secreting tumor. This is a tumor that secretes ACTH, but is not found in an endocrine organ.

A physician can diagnose the tumor type using the results from two diagnostic measures. The physician will check for the presence or absence of hyperpigmentation. And also, the physician will administer dexamethasone, which is a cortisol analogue that acts on endocrine organs. The results of these diagnostics, plus well-developed critical thinking skills, allow the physician to make a diagnosis. Try it yourself: What is the likely cause of Cushing syndrome in a patient who exhibits hyperpigmentation, and who is unresponsive to dexamethasone?

❶ How is it possible that all three types of tumors can lead to the same set of classic Cushing syndrome symptoms?

In Cushing syndrome patients, symptoms can be resolved in a matter of days by administering cortisol receptor blockers. If cortisol receptor blockers are successful in alleviating symptoms, the symptoms are necessarily caused by cortisol. Therefore, it can be inferred that cortisol, secreted by the cortex of the adrenal glands, mediates the psychological symptoms associated with Cushing syndrome. In fact, this syndrome is specifically defined as a collection of signs and symptoms due to prolonged exposure to cortisol.

The three tumor locations associated with Cushing syndrome are in the anterior pituitary, the adrenal cortex, or in an ectopic tumor site that secretes ACTH. ACTH is a hormone that acts on the cortex to cause the release of cortisol. Thus, via ACTH production or via direct cortisol production, all three of these tumor locations (indicated in the figure by the tumor-like 'bumps') increase the amount of cortisol available in the body, causing Cushing syndrome.

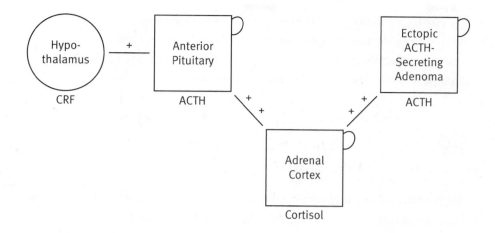

2 **Why wouldn't dexamethasone rule out a tumor of the adrenal cortex or an ectopic ACTH-secreting tumor?**

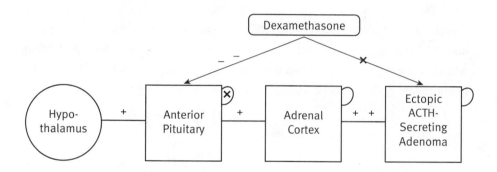

The prompt states that dexamethasone is a cortisol analogue that acts on endocrine organs. In the HPA axis, the hypothalamus and anterior pituitary respond to cortisol by downregulating CRF and ACTH production as part of a negative feedback loop. Dexamethasone would have a similar effect, and would cause tissue of the anterior pituitary, including an anterior pituitary tumor, to downregulate production of ACTH and thus cortisol production as well. If the tumor were of the anterior pituitary, there would be a response to the administration of dexamethasone. An ectopic ACTH-secreting adenoma, such as a lung tumor, does not express cortisol receptors, since the lung tissue is not supposed to be responsive to or induce cortisol in the first place. The same can be said for the adrenal cortex itself, which also would not express significant quantities of negative-feedback related cortisol receptors.

The prompt states that the patient is nonresponsive to dexamethasone, so the tumor must be in tissue that does not normally respond to cortisol. Neither the adrenal cortex nor an ectopic tumor would be expected to have a negative feedback effect in response to cortisol, so they won't respond to dexamethasone either.

Takeaways

Negative feedback systems are common throughout the endocrine system. Many of these systems can be organized into axes, including the hypothalamic–pituitary–adrenal, hypothalamic–pituiary–thyroid, and hypothalamic–pituitary–ovarian, –testicular, or –gonadal axes.

Things to Watch Out For

Proper diagnosis of an endocrinopathy usually requires measurement of at least two different hormones or analysis of at least two different organs. This allows determination of the site of the problem. In this question, a high concentration of cortisol could be due to a problem in the brain or in the adrenal glands; measurement of ACTH would distinguish between these two sites of pathology.

3 Why does hyperpigmentation have to be due to elevated ACTH levels?

The prompt states that hyperpigmentation is the only symptom of Cushing syndrome not resolved by cortisol receptor blockers. This means that the symptom of hyperpigmentation must be caused by a source outside of cortisol, but involved in the same pathway as it is a symptom in some patients. As Cushing syndrome is not caused by tumors of the hypothalamus and cannot be linked to CRF, the only remaining hormone that could cause hyperpigmentation is ACTH.

In an adrenal cortical tumor, the only hormone related to Cushing syndrome that is produced is cortisol, meaning a tumor of the adrenal cortex could not cause hyperpigmentation. A tumor that secreted ACTH, however, would continue to overproduce ACTH even in the presence of cortisol receptor blockers, which would explain why the hyperpigmentation would not respond to the drug treatment. Therefore, the tumor in this system must be capable of secreting ACTH.

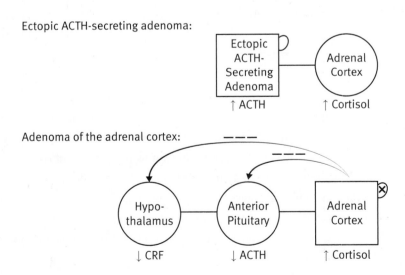

A patient with hyperpigmentation unresponsive to dexamethasone likely suffers from ectopic Cushing syndrome.

Related Questions

1. A pituitary tumor compresses the stalk and causes a cessation of ACTH secretion. What effects would this lack of ACTH secretion have?

2. The posterior pituitary gland is also known as the neurohypophysis. Based on the relationship to other brain structures, why does this name make sense?

3. The pancreas is said to have both exocrine and endocrine functions. What is the difference between an exocrine and an endocrine organ?

⊟ The Respiratory System

The volume of the lungs that does not participate in gas exchange is considered physiological dead space. There are two types of dead space that are seen at rest: anatomical and alveolar. Anatomical dead space includes the conducting areas, such as the mouth and trachea, where oxygen enters the respiratory system but does not contact alveoli. Alveolar dead space is the area in the alveoli that does contact air but lacks sufficient circulation to participate in gas exchange. How can physiological dead space be reduced?

1 What are the characteristics of the different types of dead space?

Anatomical dead space refers to the air that remains in the conducting pathways—the mouth and trachea—with every breath. Because the size and length of the mouth and trachea are set and relatively unchangeable for a given individual, it is unlikely that physiological dead space can be decreased much through changes in the anatomical dead space.

Alveolar dead space consists of alveoli that contact air but do not participate in gas exchange. Because the alveoli are normal, they are capable of participating in gas exchange under the right conditions; therefore, alveolar dead space can be reduced.

2 How does gas exchange occur at the lungs and tissues?

In the normal lung, oxygen will diffuse from alveolar air into the pulmonary capillaries. When the partial pressure of O_2 in the alveolar air and capillary blood equilibrate, the diffusion stops. Normally, this occurs before the blood in the pulmonary capillary exits the lungs and is considered perfusion-limited gas exchange. This O_2 is bound to hemoglobin and is taken and released to the tissues. CO_2 is produced by the tissues and diffuses into capillary blood, where it is carried to the lungs primarily as HCO_3^-. At the lungs, the reaction is reversed and CO_2 is exhaled:

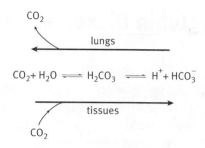

$$CO_2 + H_2O \rightleftharpoons H_2CO_3 \rightleftharpoons H^+ + HCO_3^-$$

3 Why do some alveoli not participate in gas exchange?

There is not sufficient bloodflow through the capillaries of these dead space alveoli to induce them to participate in gas exchange. There must be bloodflow to a given alveolus for gas exchange to occur in that alveolus.

4 How could bloodflow through the lungs be increased?

If pulmonary bloodflow were increased, then more alveoli would be perfused with blood and would therefore participate in gas exchange. Increasing pulmonary bloodflow would require increasing the output of the right ventricle. Cardiac output increases during exercise because there is an increased heart rate and increased venous return due to skeletal muscle activity. Therefore, exercise would increase the amount of pulmonary bloodflow. This increased flow of blood through the lungs would recruit more alveoli for gas exchange and, therefore, reduce alveolar and physiological dead space.

Pulmonary bloodflow could also be increased through the administration of vasodilators, which open the blood vessels, lowering vascular resistance and increasing flow. Certain vasodilators are particularly active in the lungs.

Related Questions

1. What is the result if bloodflow to the left lung is completely blocked by a pulmonary embolus?

2. If an area of the lung is not ventilated due to an obstruction of the airways, what is the partial pressure of oxygen (P_aO_2) in the pulmonary capillaries of that area?

3. At what point will the diffusion of oxygen from the alveolus to the capillary stop?

Takeaways

The respiratory system is intimately linked to the circulatory system. Oxygen is delivered to tissues, and CO_2 is removed from tissues and ultimately removed from the body through exchange from the pulmonary capillaries into the alveolar space.

Things to Watch Out For

Under resting conditions, alveolar oxygen equilibrates with the blood in the pulmonary capillaries. This is considered a perfusion-limited exchange. Under conditions of exercise, the partial pressures of oxygen do not necessarily equilibrate along the length of the pulmonary capillary and a partial pressure gradient is maintained even at the venous end of the capillary.

Key Concepts

Biology Chapter 7
Oxyhemoglobin dissociation curve
Carbon monoxide
Hemoglobin
Left and right shifts

⑤ Oxyhemoglobin Dissociation Curve

Carbon monoxide (CO) binding to hemoglobin occurs in competition with oxygen (O_2) binding to hemoglobin; hemoglobin's affinity for CO is over 200 times its affinity for O_2. However, the binding of CO at one site increases the affinity for O_2 at the remaining sites. Draw the oxyhemoglobin dissociation curve for CO poisoning, measuring hemoglobin oxygen content (in units of % hemoglobin saturation) on the vertical axis.

1 **What two factors contribute to the sigmoidal shape of the oxyhemoglobin dissociation curve?**

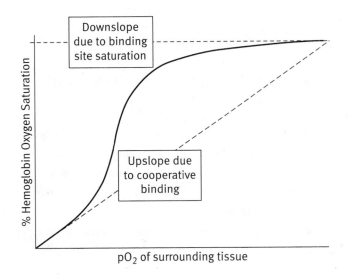

Oxygen binds cooperatively to hemoglobin, meaning that the binding of the first oxygen molecule facilitates the binding of the next one, and so on. This binding trend causes the curve to increase slope once oxygen has begun to bind, as subsequent binding is being facilitated. The asymptote at the top of the sigmoidal shape is caused by saturation of the hemoglobin in the body. When all hemoglobin has bound four oxygen atoms, no more binding can occur, independent from pressure of oxygen available. These two factors cause the shift toward a steeper curve at the low pO_2 end of the graph, and then the leveling off at the high pO_2 end of the graph.

2 How would carbon monoxide impact the curve?

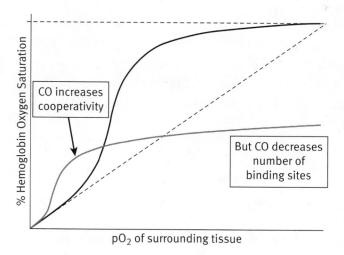

The binding of CO occupies one of the binding sites for O_2 on hemoglobin, which functionally lowers the maximum number of O_2 that can bind to hemoglobin. This decreased total binding capacity lowers the asymptotic maximum of the curve. However, the binding of CO also increases hemoglobin's affinity for O_2 at the remaining sites. Any physiological factor that increases affinity of hemoglobin for oxygen has the effect of shifting the curve to the left. Thus, CO shifts the curve toward the left. It is this left shift that makes CO poisoning so deadly; with CO bound to hemoglobin, the O_2 molecules are bound so tightly that they cannot be offloaded at the tissues, and tissue hypoxia occurs.

Related Questions

1. How would exercise affect the oxyhemoglobin dissociation curve?

2. What type of compensatory reactions would a high PCO_2 cause?

3. What would the oxyhemoglobin dissociation curve look like in a patient with metabolic alkalosis?

Takeaways

Be familiar with the concepts of PO_2 and PCO_2, what factors change their values, and how they are represented graphically.

Things to Watch Out For

Right shifts in the dissociation curve (Bohr effect) are more commonly seen, but be prepared for the factors that can cause the curve to shift to the left, including decreased PCO_2, increased pH, decreased temperature, fetal hemoglobin, and certain pathological states—such as carbon monoxide poisoning.

E Immune Function

X-linked severe combined immunodeficiency (SCID) occurs due to a mutation in a protein necessary for the receptors for interleukins IL-2, IL-4, IL-7, IL-9, and IL-15. This results in the absence or near absence of T-cells and natural killer (NK) cells. B-cells are present but are nonfunctioning in these individuals. Why, despite their presence, are B-cells nonfunctional in those with SCID?

1 How could this question be simplified?

The question stem states that severe combined immunodeficiency (SCID) results when there is a mutation in a protein required for the receptors for various interleukins. This results in a lack of T-cells and natural killer (NK) cells. B-cells are still present but are nonfunctional. Given this information, this question is really asking about how a receptor mutation will cause a lack of some cell types (T-cells and NK cells), while producing nonfunctional B-cells.

2 What is the normal purpose of interleukins?

Interleukins are the language of communication in the immune system. Interleukins direct the proliferation and maturation of lymphocytes including T-cells and B-cells. In addition, portions of the innate immune system are also coordinated by interactions with interleukins. Interleukins are known to activate macrophages, NK cells, and B-cells.

3 How would a mutation in an interleukin receptor affect immune signaling?

Communication between cells often requires interactions between some soluble molecule and a receptor. Lack of a functioning receptor often has the same effect as a lack of the effector molecule.

The question stem states that there is a mutation in a protein required for interleukin receptors. This mutation affects the function of multiple interleukins. The dysfunction of this interleukin receptor results in defective immune system signaling. If the immune system cannot signal between cells properly, then numerous processes may be affected. Interleukins are part of the system that induces proliferation of T-cells and NK cells. In addition, without effective interleukin signaling, T-cells that are produced may not be able to migrate to the thymus for maturation. A lack of mature T-cells will result in the inability of T-cells to signal to B-cells. If there is no signaling to B-cells, then plasma and memory cells will not form, and antibody production will not occur. Therefore, there will be little to no humoral protection against antigens in these individuals.

④ Why might B-cells be nonfunctional in patients with SCID?

To summarize the information previously considered, lack of a functioning interleukin receptor results in a near complete loss of function of the adaptive immune system. T-cells and NK cells do not mature, and B-cells are not able to function. If the same protein is involved in multiple receptor types, then mutations of this protein will affect multiple interleukin receptors. The resulting phenotype is a nonfunctioning adaptive immune system. T-cells and NK cells are unable to proliferate, and signaling to B-cells is severely limited.

Related Questions

1. During pregnancy, some immunoglobulins are transferred across the placenta. After birth, breast milk also contains antibodies that provide the infant with reduced susceptibility to infection. What is a potential disadvantage of this type of immunity?

2. Allergic reactions, such as seasonal allergies, rely on the cross-linking of immunoglobulins that are already on the surface on mast cells. Why does the first exposure to an allergen rarely result in an immune response?

3. Respiratory droplets, created during sneezing or coughing, can carry bacteria and viruses. What type of immunity is represented by the cilia of the respiratory passages that carry these viruses toward the oropharynx to be swallowed or expelled?

Takeaways

Because of the interdependencies of the immune system, dysfunction of one arm of the immune system may cause dysfunction of other arms of the immune system. In this example, a lack of functional T-cells leads to a lack of functional B-cells.

Things to Watch Out For

Many of the questions focusing on the immune system on the MCAT focus on the crosstalk of this system. Most important among these forms of communication are those between the innate and adaptive immune systems and those between T- and B-cells.

⊟ The Digestive System

During the gastric phase of digestion, the presence of food in the stomach, particularly amino acids and peptides, causes G cells to secrete gastrin, which in turn stimulates parietal cells. Gastrin secretion is normally inhibited once acidic chyme, with a pH less than 3, reaches the duodenum. What would occur if a tumor secreted excess gastrin in an unregulated fashion?

1 What is the role of gastrin in the stomach?

According to the question stem, gastrin stimulates parietal cells when food is present in the stomach. Parietal cells secrete hydrochloric acid, and therefore gastrin is a physiological agonist of HCl secretion. Once the chyme reaches a certain acidity (pH < 3) and moves into the small intestine, gastrin secretion is inhibited and therefore HCl secretion is decreased.

2 What is the role of hydrochloric acid in the stomach?

In the stomach, HCl is necessary for the proper function of *pepsin* because low pH promotes activation of *pepsinogen* to pepsin; this enzyme is also most active in the pH range of 1–3.

3 What happens when acidic chyme reaches the small intestine?

Once the chyme moves into the small intestine, the pH needs to be increased to reach the optimal pH for pancreatic proteases and lipases (around 8.5). Therefore, gastrin release is inhibited, and the pancreas is stimulated to secrete bicarbonate into the duodenum to neutralize the acid. The pancreas also releases hydrolytic enzymes, such as *amylase*, *trypsinogen*, *chymotrypsinogen*, *carboxypeptidases* A and B, and pancreatic lipase.

4 What would be the effect of a gastrin-secreting tumor?

A gastrin-secreting tumor will secrete gastrin at all times and will not be inhibited by normal feedback mechanisms, such as the presence of chyme in the small intestine. This gastrin will continually stimulate parietal cells to produce HCl. This excess of acid will move with the chyme into the small intestine. Normal amounts of bicarbonate will be released; however, this is not enough to neutralize such an excess of HCl.

This condition—caused by a gastrin-secreting tumor—is called *Zollinger–Ellison syndrome*. The tumor is usually located in the pancreatic islets of Langerhans.

5 How would an acidic environment inside the small intestine affect function?

Pancreatic enzymes require a more alkaline environment than gastric enzymes for optimal activity. If the environment in the small intestine is too acidic, then pancreatic enzymes will be unable to function normally. While proteins and carbohydrates are partially digested before they reach the small intestine, fats have very little digestion until they reach the duodenum. If *pancreatic lipase* is unable to function due to an excessively acidic environment, it will not be able to digest lipids. This will result in the malabsorption of lipids, which leads to excretion of fats in the feces, also known as *steatorrhea*. The excess acid can also cause gastric and duodenal ulcers, in which the mucosal lining is worn away and acid can damage the underlying epithelial cells.

Takeaways

The effects of excess acid on the digestive system include the direct effects of acid on the tissues (gastric and duodenal ulcers) as well as biochemical effects on the digestive enzymes (denaturation and decreased activity).

Things to Watch Out For

An acidic duodenal pH will affect the function of all of the pancreatic enzymes, but some protein absorption can still occur because proteins are partially digested in the stomach. Carbohydrate digestion does not rely heavily on pancreatic secretions; absorption of some carbohydrates can still occur because they begin digestion in the mouth and continue to be broken down by disaccharidases in the intestinal brush border.

Related Questions

1. A patient with a peptic ulcer takes a large dose of antacid. This would primarily reduce the activity of which digestive enzyme?

2. Pancreatic ductal cells secrete bicarbonate, which is moved into the intestinal lumen. What would be the physiological results if these ductal cells were destroyed by an autoimmune disorder?

3. Pancreatitis is a disease that prevents the pancreas from being able to produce adequate amounts of lipase. What will be the physiological result of this component of the disease?

S Kidney Function

Acesulfame potassium is a calorie-free artificial sweetener used to sweeten many popular diet sodas. Within the GI tract, acesulfame potassium is rapidly and completely absorbed; however, it is neither stored nor metabolized within the body, and is instead filtered and excreted by the kidney. A student discovers that drinking diet caffeinated soda, sweetened with acesulfame potassium, results in a significant increase in urine volume and frequency. What are the physiological factors driving this phenomenon?

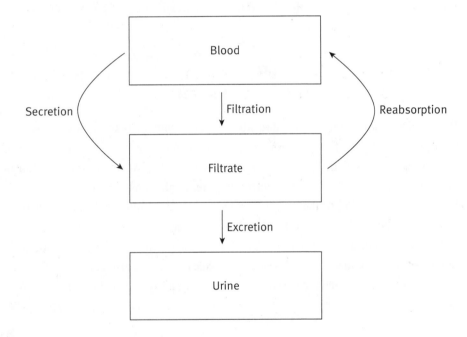

1 Why might a reduction in reabsorption increase urinary output?

Reabsorption is the process by which fluid or solutes are taken from the filtrate (inside the tubules of the nephrons) and put back into the blood. Thus, if reabsorption is inhibited or reduced, less of the filtrate will return to the blood and more will be excreted via the urine. Similar results are seen with an increase in secretion. Secretion is the opposite of reabsorption and plays a critical role in maintaining blood pH, potassium concentrations in the blood, and nitrogenous waste concentrations in the filtrate. Since water follows the movement of solutes, increasing secretion has the effect of increasing urinary output.

② **Why does caffeine's inhibition of antidiuretic hormone increase urine volume?**

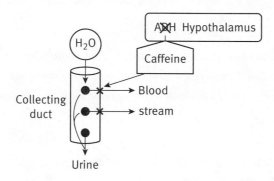

Antidiuretic hormone (ADH) decreases urinary volume. It works directly on the collecting duct by increasing permeability to water; thus, a decrease in ADH levels will lead to a decrease in water reabsorption. The net effect of ADH is therefore a decrease in urinary output. ADH secretion is normally triggered by a sustained increase in plasma osmolarity. If ADH is not released or cannot activate its receptor, water will not be reabsorbed and the plasma osmolarity will remain high. If the water remains in the urine, it increases urinary volume and frequency.

③ **By what mechanism does ingesting a large volume of soda increase urine volume?**

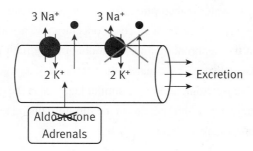

Ingesting a large volume of soda increases arterial pressure, which leads to a decrease in renin and aldosterone secretion and ultimately a decrease in water reabsorption. This effect is consistent with the fact that the body needs a certain amount of water to maintain homeostasis. Therefore, a large urinary output would naturally follow a large water intake.

Recall that renin is released by the kidneys in response to low blood pressure. Renin activates angiotensinogen to angiotensin I, which is converted to angiotensin II by

Takeaways

Questions related to the nephron appear in varying contexts. The key is to isolate the relevant component of kidney function being tested (filtration, secretion, and/or reabsorption) and then to tease apart which particular components of that function (such as filtrate osmolarity or hormonal control) are being affected.

Things to Watch Out For

Under most circumstances, the kidney produces urine that is hypertonic to the blood, but in the situation described here, the urine produced is likely to be hypotonic to the blood. Alcohol consumption produces similar physiological effects. However, frequent urination does not always mean that the urine is hypotonic to the blood. Patients who excrete protein in their urine (filtration failure) have low blood osmolarities, and thus have a low level of water reabsorption. Therefore, the urine produced may still be hypertonic to the blood.

angiotensin-converting enzyme (ACE) in the lungs. Angiotensin II promotes secretion of aldosterone from the adrenal cortex. Recall that aldosterone increases sodium reabsorption in the distal convoluted tubule and collecting duct, and because water follows sodium from the filtrate into the blood, aldosterone also increases water reabsorption.

If renin and aldosterone secretion is reduced, then sodium and water will not be reabsorbed in the distal convoluted tubule and collecting duct, and will therefore be excreted in the urine. The increase in water increases urinary volume and frequency.

How does acesulfame K contribute to increasing urinary output?

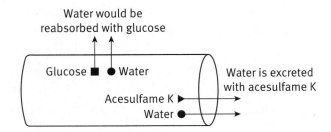

Diet sodas substitute acesulfame potassium for sugar, but unlike glucose, the acesulfame potassium cannot be reabsorbed back into the blood from the nephron—hence the high filtrate osmolarity. Even if you did not know that acesulfame potassium cannot be reabsorbed from the nephron, you should recognize the role that filtrate osmolarity can play in water reabsorption. Recall that in diabetes mellitus, a similar mechanism is at play: due to the high glucose concentration, not all of the sugar can be reabsorbed from the nephron, leading to an abnormally high filtrate concentration, less water reabsorption, and ultimately the excretion of a larger volume of urine that also contains glucose. Something very similar happens in the case of acesulfame potassium: acesulfame potassium remains in the filtrate, causing less reabsorption of water, and increasing urinary volume as a result!

Related Questions

1. A patient has been found to have insufficient levels of antidiuretic hormone. What symptoms would this individual have?

2. Diabetics who fail to take insulin experience dehydration. What are the physiological factors driving this phenomenon?

3. A patient with renal failure has nephrons that lack the ability to actively secrete or reabsorb any substances. What functions might the kidney still be able to perform?

E Muscle Contraction

> What process during muscle contraction is likely to be affected by a deficiency of parathyroid hormone?

1 What is the function of parathyroid hormone?

Parathyroid hormone (PTH) is involved in calcium homeostasis. Release of parathyroid hormone results in an increased calcium concentration in the blood. Specifically, PTH decreases excretion of calcium by the kidneys, increases absorption of calcium in the gut (via vitamin D), and increases bone resorption, thereby freeing up calcium. PTH is subject to feedback inhibition—as levels of plasma calcium rise, PTH secretion is decreased.

PTH also affects phosphorus homeostasis by resorbing phosphate from bone and reducing reabsorption of phosphate in the kidney (thus promoting its excretion in the urine). PTH also promotes the activation of vitamin D, which is required for the absorption of calcium and phosphate in the gut. For this question, we will focus primarily on the role of PTH in calcium homeostasis.

2 What are the effects of a parathyroid hormone deficiency?

If parathyroid hormone is not released, then all of these mechanisms will be downregulated. Calcium excretion in the kidneys will be decreased, less calcium would be absorbed from the gut, and bone resorption would decrease. This could lead to hypocalcemia, or low concentrations of calcium in the blood.

3 What is the role of calcium in muscle contraction?

During muscle contraction, depolarization of the muscle cells results in the opening of calcium channels in the sarcolemma. Calcium then binds to troponin, causing a conformational change in tropomyosin, which exposes the myosin-binding sites on actin.

What process is affected by low calcium levels?

If prolonged enough, low calcium levels in the blood can lead to low calcium levels within the sarcoplasmic reticulum of muscle cells. When stimulated, these muscle cells will have a smaller efflux of calcium from the sarcoplasmic reticulum, which will affect the binding of calcium to troponin.

Under normal circumstances, it is the binding of calcium to troponin that causes a conformational change in tropomyosin, exposing the myosin-binding sites on actin and permitting cross-bridge formation for muscle contraction. Therefore, a decreased calcium concentration in the muscle cell will lead to less cross-bridge formation and a weaker muscle contraction. Because each muscle fiber contains numerous molecules of troponin, this may also result in an inability to recruit all of the muscle fibers in an organized, efficient fashion.

Related Questions

1. An autoimmune disease produces antibodies against the acetylcholine receptor at the neuromuscular junction. Which step in muscle contraction is likely to be affected most directly by this disease?

2. Which division of the sarcomere does not change in size during muscle contraction?

3. A student is attempting to identify types of muscle tissue by microscopic examination. A sample contains cells with one or two nuclei and intercalated discs between the cells. The overall appearance is striated. What muscle type is likely included in this sample?

Takeaways

The intracellular environment of the muscle cell can be impacted by its extracellular environment. Prolonged changes in blood concentrations of various electrolytes can impact the internal environment of muscle (and other) cells, leading to changes in their physiology.

Things to Watch Out For

This question was very specific. You may know that one of the symptoms of hypocalcemia is tetany, but that phenomenon is related to changes in permeability of neurons, rather than muscle contraction itself. When a question is very specific, provide a specific answer related to the theme of the question stem.

Key Concepts

Biology Chapter 12
Hardy–Weinberg equilibrium
$p^2 + 2pq + q^2 = 1$
Allele frequencies
Genotype frequencies

Ⓢ Hardy–Weinberg Equilibrium

> Gigantism is coded for by a recessive allele. The dominant allele for the same gene codes for the normal phenotype. In an isolated geographic area, 9 people out of a sample of 10,000 were found to have gigantism, whereas the rest had normal phenotypes. Assuming Hardy–Weinberg equilibrium, calculate the frequency of the recessive and dominant alleles as well as the number of heterozygotes in the population.

Hardy–Weinberg Equilibrium: Eq 1: $p + q = 1$ Eq 2: $p^2 + 2pq + q^2 = 1$

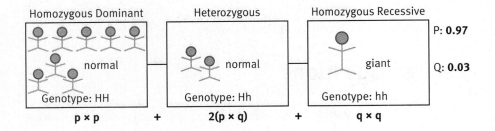

1   How can we successfully calculate the frequency of the recessive allele to be 3%?

You can use the Hardy–Weinberg equilibrium to calculate the frequency of the recessive allele for gigantism. Gigantism will only emerge with a homozygous recessive genotype, which you can represent as hh. In the Hardy–Weinberg equation, the recessive genotype frequency is depicted as q^2. By taking the square root of q^2, you get the frequency of the recessive gigantism allele, q, which is 0.03:

$$\text{Gigantism} = \text{homozygous recessive} = \text{gg} = q^2$$

$$q^2 = \frac{9}{10000} = 0.0009$$

$$q = 0.03$$

recessive allele frequency = 3%

K

2 Why is the frequency of the dominant allele 97%?

The sum of the frequency of the dominant allele, p, and the frequency of the recessive allele, q, equals 1. To solve for the frequency of the dominant allele, simply subtract the recessive allele frequency from 1:

$$p + q = 1$$
$$p = 1 - q$$
$$= 1 - 0.03$$
$$= 0.97$$

dominant allele frequency (p) = 97%

3 Why must 2pq be used to solve for heterozygote frequency?

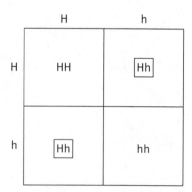

Genotypic Frequency Derivation

Takeaways

Remember both Hardy–Weinberg equations: $p + q = 1$ and $p^2 + 2pq + q^2 = 1$. p is the frequency of the dominant allele; q is the frequency of the recessive allele. p^2 is the frequency of the homozygous dominant genotype, $2pq$ is the frequency of the heterozygous genotype, and q^2 is the frequency of the homozygous recessive genotype.

Things to Watch Out For

There are five circumstances in which the Hardy–Weinberg equations may fail to apply. These are: mutations, migration, genetic drift, nonrandom mating, and natural selection. The Hardy–Weinberg equations refer to populations that are stable and *not* evolving.

The second Hardy–Weinberg equation uses a two-allele system and a double-heterozygote cross in order to predict estimated frequency. Allele frequencies can be predicted using a Punnett square. When the population of all possible offspring is summed, the entire population can be predicted, and that value can be set equal to 1 as seen with the second iteration of the Hardy–Weinberg equation. You can use $2pq$ from this expression to calculate frequency of heterozygotes.

$$[HH] + [Hh] + [Hh] + [hh] = p^2 + 2pq + q^2 = 1$$

$$2pq = \text{frequency of heterozygotes}$$

$$Hh = 2pq = 2 \times 0.97 \times 0.03$$

$$= 0.058 = 5.8\%$$

To find the size of the heterozygote population, multiply the frequency by the total size of the population:

$$0.058 \times 10,000 = 580 \text{ people}$$

Related Questions

1. Suppose a similar survey was done in another area. This time, 90 people with gigantism were found from a survey of 200,000 people. Calculate the same parameters with the new survey.

2. An allele f occurs with a frequency of 0.8 in a wolfpack population. Determine the frequencies of the genotypes FF, Ff, and ff.

3. If the frequency of homozygous recessive individuals for a certain autosomal recessive disease is 16 percent, determine the percentage of phenotypically normal individuals.

Solutions to Related Questions

1. Prokaryotic Genetics

1. In this question, we must think about the characteristics required to induce the transcription of the DNA encoded on the plasmid. First, in order for bacteria to transcribe the genetic sequence, an appropriate promoter must be present. In addition, the genetic material to be transcribed must be close enough to the promoter to facilitate effective transcription. In order to ensure efficient conversion of all of the bacteria in a colony to bacteria able to produce this protein, the plasmid should also encode a sex factor so that the plasmid can be efficiently transferred between cells.

2. The fact that the same bacterial genetic sequence is transferred each time this bacteriophage infects a bacterium implies that the process of transduction is not random in this case. This is likely the case if the phage genome integrates into the bacterial genome and, when the phage genome is later released from the bacterial genome, it takes a small segment of bacterial DNA with it. This implies that the bacteriophage must enter the lysogenic cycle. To replicate and subsequently infect other bacteria, the bacteriophage will also have to reenter the lytic cycle at some point.

3. Sometimes, the sex factor becomes incorporated into the bacterial genome. This information is generally carried on a plasmid, but, by the process of transformation, it may become part of the bacterial genome itself. When this occurs, the entire genome will be replicated and the bacterium will attempt to transfer the entire genome during conjugation. The conjugation bridge is usually not stable enough to permit transfer of the entire bacterial genome, but sizable portions of it are often still transferred. This results in high frequency recombination. Cells that have undergone this particular change are known as Hfr or high frequency of recombination cells.

2. The Menstrual Cycle

1. FSH is inhibited early in the follicular phase to prevent the development of multiple follicles. While some FSH is necessary for the development of a follicle at all, the body prevents the formation of many follicles by negative feedback of estrogen on the hypothalamus and anterior pituitary. Later in the follicular phase, FSH levels rise in parallel with LH levels as estrogen's feedback mechanism switches from negative feedback to positive feedback.

2. Progesterone levels are very low during the follicular phase of the menstrual cycle, and do not begin to rise until the luteal phase. The corpus luteum, which does not exist until the secondary oocyte has been released from the ovary, secretes progesterone to develop and maintain the endometrium. Thus, progesterone levels rise during the luteal phase. It is the drop in the level of progesterone that ultimately stimulates menses, which—by definition—begins on day 0 of the cycle.

3. Ovulation can be prevented biochemically by manipulating the hormones involved in the menstrual cycle. Birth control pills, or oral contraceptive pills (OCPs), keep estrogen and progesterone levels elevated so that FSH and LH are constantly inhibited. Because of this inhibition, a follicle does not develop and the LH surge does not occur, thereby preventing ovulation.

3. Stages of Embryogenesis

1. Trophoblasts are the cells that form the placenta. In formation of the placenta, these cells must invade the endometrium to create the interface between the mother and the developing embryo. This means that this cell type is particularly inclined toward rapid division and invasion of foreign tissues. Thus, if a malignancy develops, the ability of this cell to invade blood vessels and spread (metastasize) is often more pronounced than with other cell types.

2. The mesoderm is the middle germ layer within the gastrula. The mesoderm gives rise to the musculoskeletal system, the circulatory system, much of the excretory system, the gonads, the muscular and connective tissue layers of the digestive and respiratory systems, and the adrenal cortex. Thus, a mutation that occurs early in the process of development will have far-reaching consequences, affecting each of these organ systems.

3. Indeterminate cleavage means that the cells are still capable of developing into any cell type, and could even develop into entire organisms. If a zygote undergoes indeterminate cleavage, and the cells actually separate from each other, then twinning may occur. Because these twins arose from the same zygote, they would be considered monozygotic or identical twins. Depending on when during development the organisms split from each other, the resulting twins could be dichorionic/diamniotic, monochorionic/diamniotic, or monochorionic/monoamniotic.

4. Action Potentials

1. The opening of the sodium channel brings the membrane potential closer to sodium's electrochemical equilibrium potential. The longer the channel is open, the closer the membrane potential will get to this equilibrium potential. Therefore, sodium is closest to its electrochemical equilibrium right before the sodium channels are inactivated. This corresponds to the highest point of the action potential in the graph—the point right between regions II and III, around +35 mV (the actual equilibrium potential for sodium is close to +60 mV).

2. An action potential is an all-or-nothing response, so the speed and magnitude of an individual action potential cannot be altered. Rather, the magnitude of a stimulus is encoded by the frequency of action potential firing. A lower-magnitude stimulus elicits a lower firing frequency; a higher-magnitude stimulus elicits a higher firing frequency.

3. When a membrane is hyperpolarized, it is further from the threshold potential and would require a larger stimulus to create an action potential. Therefore, hyperpolarizing a membrane would be a way to inhibit an action potential. Action potentials are also inhibited during the absolute refractory period, which occurs during the action potential itself. Finally, blockage or inactivation of the sodium channel will prevent the depolarization phase of the action potential, thereby inhibiting the rest of the action potential from occurring.

5. The Endocrine System

1. ACTH, or adrenocorticotropic hormone, is one of the tropic hormones released by the anterior pituitary. The hypothalamus releases corticotropin-releasing factor (CRF), which triggers the release of ACTH from the anterior pituitary. Then, ACTH stimulates cortisol secretion from the adrenal cortex. Cortisol is one of the hormones that aids in wakefulness as well as aiding in the maintenance of blood pressure. Lack of cortisol is likely to result in fatigue and low blood pressure.

2. The posterior pituitary is best thought of as an extension of the hypothalamus. The bodies of the nerves of the posterior pituitary are actually located in the hypothalamus, which sends axons down the length of the pituitary stalk into the region of the posterior pituitary. The two main hormones released from the posterior pituitary, antidiuretic hormone (ADH or vasopressin) and oxytocin, are made in the hypothalamus and stored in the nerve terminals of the posterior pituitary. Thus, *neurohypophysis* is an appropriate name for the posterior pituitary as this organ is simply an extension of certain hypothalamic neurons.

3. An exocrine organ is one that secretes its products into ducts, which empty onto an epithelial surface. Examples of exocrine organs include lacrimal glands, sweat glands, salivary glands, mammary glands, and the digestive portions of the pancreas. Endocrine organs secrete hormones into the bloodstream, where the hormones travel to distant target tissues. Examples of endocrine organs include the hypothalamus, pituitary gland, thyroid, parathyroid glands, pancreas, adrenal glands, and gonads.

6. The Respiratory System

1. If respiration is occurring normally but there is no bloodflow to the left lung, then there is no gas exchange occurring in the left lung. If no oxygen is being diffused from the air into the blood and no CO_2 is being released, then the P_AO_2 of the alveoli will equal the PO_2 of inspired air, around 100 mmHg O_2.

2. If there is an airway obstruction but bloodflow to the lung is normal, then there is no gas exchange. The blood that flows into the lung will have the same PO_2 and PCO_2 as systemic venous blood. Because no gas exchange occurs, this value will remain the same for blood that exits the lung and throughout the pulmonary capillaries in that area. P_VO_2 is approximately 40 mmHg and P_VCO_2 is approximately 47 mmHg.

3. When the P_AO_2 of inspired air equals the P_aO_2 of capillary blood, then diffusion of oxygen from the alveolus to the capillary will stop. Note that in physiology, *A* stands for alveolar, whereas *a* stands for arterial.

7. Oxyhemoglobin Dissociation Curve

1. Exercise will lead to an increase in body temperature as well as an increase in CO_2 production by the tissues. Through the bicarbonate buffer system, the excess CO_2 generates protons, lowering the pH. The pH may be further lowered if anaerobic metabolism is occurring, generating lactic acid. These factors cause the Bohr effect—a shift in the oxyhemoglobin dissociation curve to the right.

2. A high PCO_2 will lead to an increase in ventilation rate. This allows the body to exhale additional carbon dioxide, bringing the body back to a normal PCO_2. In prolonged cases of hypercarbia (high concentrations of carbon dioxide in the blood), the kidney may begin excreting additional acid, shifting the bicarbonate buffer system to lower the concentration of carbon dioxide in the blood.

3. In metabolic alkalosis, blood pH will be increased due to the loss of H^+ ions (or production of bicarbonate ions). Increased pH shifts the oxygen dissociation curve to the left, increasing the affinity of hemoglobin for oxygen.

8. Immune Function

1. Infants can receive antibodies from their mothers across the placenta and through breast milk. This type of immunization is known as passive immunity. The infant receives IgG antibodies while *in utero*, and then receives IgA antibodies while breast-feeding. This provides some protection from immunity while the child is very young. However, this form of immunity is not permanent and wanes rapidly when exposure to these antibodies ceases. The only way to maintain ongoing protection against a particular antigen is through the creation of memory cells. Without memory cells, the infant will be susceptible to the antigen as soon as the passive immunity wanes. Thus, the primary disadvantage of this type of immunity is that it is not permanent.

2. The question stem states that the immunoglobulins required for an allergic reaction are already attached to mast cells. This means that the immunoglobulins must already be present in the body. During the first exposure, the immunoglobulins required for allergic reactions have not yet been synthesized. During the creation of immunoglobulins, the first ones created are of the IgM class. After exposure to the antigen, isotype switching occurs, which allows for the production of IgE—the main form of antibody involved in allergic reactions. It is these IgE molecules on mast cells that must cross-link in order to cause degranulation of the mast cell and the release of the inflammatory mediators that cause an allergic reaction. Thus, the reason that the first exposure does not result in an allergic reaction is that IgE is formed *after* and *in response to* this first exposure.

3. Cilia, along with mucus, line the respiratory tract. The mucus is able to entrap an offending substance, and the cilia are able to move this mucus up to the oropharynx, where it can be swallowed or expelled. Because this type of immunity is not specific to a particular antigen, it is known as innate immunity.

9. The Digestive System

1. A large dose of antacid would increase the pH of the stomach and therefore inactivate pepsin. The optimum pH for pepsin is between 1 and 3; pepsin is denatured and inactivated at pH values greater than 5.

2. If the ductal cells were destroyed and the pancreas could not secrete bicarbonate, the acidic chyme that moves into the duodenum would not be neutralized and pancreatic lipase would not be able to function. This would result in the malabsorption of lipids and steatorrhea (fatty stools). This may also leave the duodenum susceptible to ulceration.

3. Inadequate amounts of pancreatic lipase will lead to steatorrhea. If lipids are not broken down by lipase, they cannot be absorbed by the small intestine and will be excreted in the feces. Steatorrhea is often described as the formation of greasy or oily stool, which is often foul-smelling due to partial breakdown by gut bacteria. These stools often float because of their high concentration of fat.

10. Kidney Function

1. If a patient has insufficient levels of antidiuretic hormone (ADH or vasopressin), he or she would suffer symptoms of diabetes insipidus, a disease characterized by ADH deficiency or lack of response by the kidneys to ADH secretion. These symptoms include excess urine production because of the inability to reabsorb water in the collecting ducts, extreme thirst due to fluid loss, and a hyperosmolar blood plasma concentration.

2. The reason why diabetics who fail to take insulin experience dehydration is linked to their excess blood glucose. Having a higher concentration of glucose in the blood means that there will be more glucose filtered into the urine—more than can be reabsorbed in the proximal convoluted tubule. With this system overloaded, glucose continues to pass through the nephron and pulls water with it. The more water that is pulled into the nephron, the more will get excreted in the urine, thus causing dehydration.

3. Just because this patient's nephrons have lost the ability to actively secrete or reabsorb substances does not mean that they cannot passively move substances. Therefore, the passive flow of ions and water will still occur. The kidney may also be able to filter blood effectively, permitting the passage of water, ions, and dissolved biomolecules while retaining cells and proteins in the vasculature. The kidney may also be able to carry out its nonurinary functions as well, such as vitamin D activation and production of erythropoietin.

11. Muscle Contraction

1. This condition, called *myasthenia gravis*, renders acetylcholine receptors nonfunctional. Acetylcholine is the neurotransmitter used at the neuromuscular junction; thus, if acetylcholine is unable to bind, then the muscle cell will not be depolarized. If depolarization does not occur, then the sarcoplasmic reticulum will not open and calcium release will not occur. Myasthenia gravis is characterized by muscle weakness, especially after repeated stimulation of the same skeletal muscle tissue.

2. The sarcomere is divided into segments (bands or zones) depending on the filaments within that segment. The Z-lines define the boundaries of each sarcomere, and the M-line defines its middle. Because a sarcomere shortens during contraction, the Z-lines and M-lines must become closer to each other. The I-band consists only of thin filaments, whereas the H-zone consists only of thick filaments. During contraction, the thin filaments slide along the thick filaments, resulting in smaller I-bands and H-zones. The A-band contains the entire length of the thick filaments. This band remains constant in size during contraction; the actin and myosin filaments slide over one another but do not change length themselves.

3. There are two types of striated muscle: cardiac and skeletal muscle. Skeletal muscle has numerous nuclei, whereas cardiac muscle contains one or two nuclei per cell. In addition, intercalated discs contain many gap junctions that allow for rapid conduction of charge between adjacent cardiac muscle cells. Therefore, it is cardiac muscle that is described in the question stem.

12. Hardy–Weinberg Equilibrium

1. If a similar survey found 90 people with gigantism out of 200,000 people, then the frequency of the recessive allele is $\sqrt{\dfrac{90}{200{,}000}} = \sqrt{\dfrac{9}{20{,}000}} = \sqrt{\dfrac{4.5}{10{,}000}} \approx \dfrac{2.1}{100} = 0.021$. The frequency of the dominant allele is $1 - 0.021 = 0.979$. The number of heterozygotes in the population is $(2 \times 0.979 \times 0.021) \times 200{,}000 \approx 2 \times 1 \times 21 \times 200 = 8{,}400$ people (actual = 8,305).

2. If allele f occurs with a frequency of 0.8, then allele F occurs with a frequency of 0.2. Thus, the frequency of FF = $0.2 \times 0.2 = 0.04$ (4%), the frequency of Ff = $2 \times 0.2 \times 0.8 = 0.32$ (32%), and the frequency of ff = $0.8 \times 0.8 = 0.64$ (64%).

3. If the frequency of homozygous recessive individuals is 0.16 (16%), then $1 - 0.16 = 0.84$ (84%) of individuals are phenotypically normal, assuming that the phenotype coded for by the dominant allele is "normal."

4

General Chemistry

Key Concepts

General Chemistry Chapter 1

Atomic mass
Atomic weight
Mass number
Isotopes

E Atomic Mass and Weight

Following an experiment, an unknown element has several isotopes with the following abundances:

$$^{64}X = 48.89\%, 63.929 \text{ amu}$$

$$^{66}X = 27.81\%, 65.926 \text{ amu}$$

$$^{67}X = 4.11\%, 66.927 \text{ amu}$$

$$^{68}X = 18.57\%, 67.925 \text{ amu}$$

$$^{70}X = 0.62\%, 69.925 \text{ amu}$$

What is the atomic weight of the unknown element? What is the element's identity? How many protons, neutrons, and electrons does $^{70}X^+$ have?

1 What is the atomic weight of the element?

The atomic weight is the weighted average of the different atomic masses. To find a weighted average, we start by multiplying the atomic mass of each isotope by its relative abundance, given as a decimal. In this example, we are precise with the calculations, but on Test Day, you would want to round these values to simplify the arithmetic and arrive at the answer more quickly:

$$63.929 \text{ amu} \times 0.4889 = 31.2549 \text{ amu}$$

$$65.926 \text{ amu} \times 0.2781 = 18.3340 \text{ amu}$$

$$66.927 \text{ amu} \times 0.0411 = 2.7507 \text{ amu}$$

$$67.925 \text{ amu} \times 0.1857 = 12.6137 \text{ amu}$$

$$69.925 \text{ amu} \times 0.0062 = 0.4335 \text{ amu}$$

Now, add together the values. Unlike average calculations in arithmetic, we have already taken into account the relative abundances and therefore do not need to divide by n, the number of samples:

$$31.2549 + 18.3340 + 2.7507 + 12.6137 + 0.4335 = 65.3868 \text{ amu}$$

Thus, the atomic weight of this element is $65.3868 \frac{\text{amu}}{\text{atom}}$; its molar mass would be $65.3868 \frac{\text{g}}{\text{mol}}$.

2 What element has an atomic weight closest to your estimate?

By definition, the weighted average of the atomic masses is the atomic weight. This is the value listed in the Periodic Table for that element's weight. The value we calculated, $65.3868 \frac{amu}{atom}$, is closest to the atomic weight of zinc (Zn), which is listed in our Periodic Table as $65.4 \frac{amu}{atom}$.

3 How many protons, neutrons, and electrons would the +1 cation have?

Looking at the Periodic Table, the atomic number of zinc is 30. The atomic number, Z, represents the number of protons in an element and actually defines the element's identity; that is, all atoms of zinc contain 30 protons.

The mass number, A, represents the number of protons plus the number of neutrons in a given atom. The various isotopes of zinc in this question have the same number of protons, but vary in their numbers of neutrons. For zinc-70, there are $70 - 30 = 40$ neutrons.

There are equal numbers of protons and electrons in a neutral atom. In the zinc-70 cation, there will be one fewer electron, giving the atom a charge of +1. Therefore, there are $30 - 1 = 29$ electrons in the zinc-70 +1 cation.

Takeaways

The atomic number is the number of protons in an element; if the element is neutral, it is also the number of electrons. The mass number is the number of protons and neutrons combined. The atomic weight is reported on the Periodic Table and is the weighted average of the atomic masses of each isotope.

Things to Watch Out For

One of the biggest traps with math-heavy general chemistry questions is trying to make calculations precise. On the MCAT, answer choices are usually far enough apart that you can round and still get sufficiently close to identify the right answer. Avoid holding onto four decimal places, as we did here, because it will greatly increase the amount of time it takes to answer the question.

Related Questions

1. Which has a larger number of molecules—a mole of carbon dioxide or a mole of water? Which one has a larger mass?

2. The atomic weight of silicon is 28.086 amu. If the isotopes in a sample are silicon-23, 24, 25, 26, and 29, which isotope is the most abundant?

3. A nuclear detector finds 80 atoms of oxygen-16 (15.999 amu), 60 atoms of oxygen-14 (13.999 amu), and 60 atoms of oxygen-18 (17.999 amu). What is the molar mass of this sample?

E Periodic Trends

Arrange the following elements in ascending order in terms of the following periodic properties: (a) electronegativity, (b) ionization energy, and (c) atomic radius.

Rb, Fr, Mg, Ca, P, Fe, Sb, O, Cl, F

1 What are the major periodic trends?

This first step just requires us to think about how periodic trends work. For all trends, there is a general direction of increasing value. Electronegativity and first ionization energy increase as one moves up and to the right in the Periodic Table (with some exceptions). Atomic radius, on the other hand, decreases in value in that direction.

2 What would be the order of the given elements from left to right?

We'll use electronegativity as our trend of choice while ordering these elements. Begin with the least electronegative element, francium (Fr), and the most electronegative element, fluorine (F), at the ends of the spectrum:

$$Fr < \_\_ < \_\_ < \_\_ < \_\_ < \_\_ < \_\_ < \_\_ < \_\_ < F$$

Let's compare oxygen (O) and chlorine (Cl), which are next to F. To determine which is more electronegative, consider the hypochlorite ion (ClO^-). The hypochlorite ion has a charge of –1. Recall that chlorine typically attains a charge of –1 in compounds, and oxygen typically attains a charge of –2, to complete their respective octet configurations. This cannot be the case in ClO^- because adding these oxidation states gives us $-1 + (-2) = -3$, when the ion's charge is –1.

Oxygen and chlorine do not have the same electronegativity, so there must be a polar covalent bond between them. If chlorine were more electronegative, then it would take on charge of –1. To maintain the overall –1 charge of the ion, oxygen's charge would have to be 0. This cannot be the case—a polar covalent bond requires that one side be slightly positive and the other slightly negative. However, if oxygen were more electronegative, then it would take on a charge of –2. To maintain the ion's overall –1 charge, chlorine's charge would have to be +1. This makes sense—there is a polar covalent bond between the two atoms, giving oxygen a slightly negative

(−2) charge and chlorine a slightly positive (+1) charge. Therefore, oxygen must be more electronegative, and we can put these two elements into the list:

$$Fr < \_\_ < \_\_ < \_\_ < \_\_ < \_\_ < \_\_ < Cl < O < F$$

The next closest atoms to the top right corner of the Periodic Table are phosphorus (P) and antimony (Sb). Phosphorus is closer to F, O, and Cl than antimony and is more electronegative:

$$Fr < \_\_ < \_\_ < \_\_ < \_\_ < Sb < P < Cl < O < F$$

Iron (Fe) is a transition metal and is the next closest to the top right corner, so it comes next:

$$Fr < \_\_ < \_\_ < \_\_ < Fe < Sb < P < Cl < O < F$$

Finally, the remaining metals can be ranked based on their proximity to Fr. Rubidium (Rb) is just two periods above Fr so it will be just slightly more electronegative than Fr. Calcium (Ca) is the next highest, and finally magnesium (Mg). Thus, the trend for electronegativity is:

$$Fr < Rb < Ca < Mg < Fe < Sb < P < Cl < O < F$$

Recognize that ionization energy should roughly parallel the order for electronegativity. For reference, the electronegativities and ionization energies of these elements are tabulated below:

Element	Fr	Rb	Ca	Mg	Fe	Sb	P	Cl	O	F
Electronegativity (Pauling Scale)	0.7	0.82	1.00	1.31	1.83	2.05	2.19	3.16	3.44	3.98
Ionization Energy $\left(\frac{kJ}{mol}\right)$	380	403	590	738	763	834	1012	1251	1314	1681

3 What would the trend be for atomic radius?

Unlike electronegativity and first ionization energy, which increase as one moves up and to the right on the Periodic Table, atomic radius *decreases* as one moves up and to the right.

Thus if,

$$Fr < Rb < Ca < Mg < Fe < Sb < P < Cl < O < F$$

Takeaways

Electronegativity, ionization energy, and electron affinity generally increase as one moves up and to the right in the Periodic Table. Atomic radius generally decreases in the same direction.

139

is the order of increasing electronegativity and first ionization energy, the reverse order is the one in which the atomic radii increase:

$$F < O < Cl < P < Sb < Fe < Mg < Ca < Rb < Fr$$

Note that there are different definitions for atomic radius; most scientists use empirical or calculated atomic radii to make predictions about periodic trends.

Related Questions

Things to Watch Out For

Remember that these are only trends, and are not absolute. There are many exceptions to the periodic trends, especially in the transition metals. Usually, if the MCAT asks for analysis of a particular periodic trend, the question will involve elements that follow that trend as expected.

1. Rank the following elements or ions from smallest to largest radius: B^{3+}, C, Be, O^{2-}

2. Which element—phosphorus or selenium—would you expect to have a higher ionization energy?

3. The first ionization energy of an element is 549.5 $\frac{kJ}{mol}$, its second ionization energy is 1064.2 $\frac{kJ}{mol}$, and its third ionization energy is 4138 $\frac{kJ}{mol}$. To which group does this element most likely belong?

High-Yield Problem-Solving Guide questions continue on the next page. ▶ ▶ ▶

Ⓢ VSEPR Theory

Penicillin is a β-lactam antibiotic, meaning that its functionality as an antibiotic depends on a central β-lactam ring. In the case of penicillin, that β-lactam ring is fused to a second, five-membered ring:

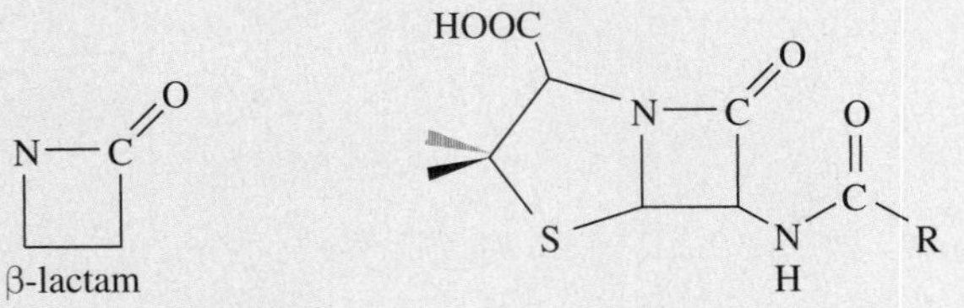

β-lactam

Penicillin

Penicillin kills bacteria by irreversibly binding to the active site of transpeptidases, a class of essential bacterial enzymes, disabling these enzymes through competitive inhibition. In this reaction, penicillin's nitrogen-carbonyl bond is broken:

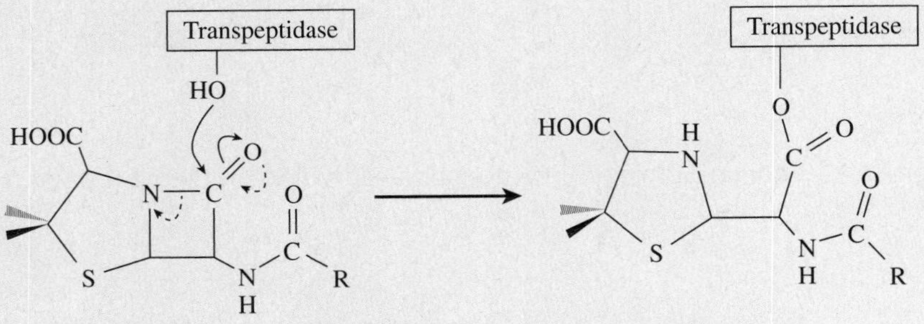

Transpeptidase

Transpeptidase

In a normal amide, this nitrogen-carbonyl bond is quite strong. However, penicillin's β-lactam structure and its fused five-membered ring both weaken penicillin's nitrogen-carbonyl bond, enabling penicillin's activity as an antibiotic. According to VSEPR theory, why do these factors weaken the nitrogen-carbonyl bond?

① **Why do the bonds around the nitrogen in a normal amide assume a trigonal planar geometry?**

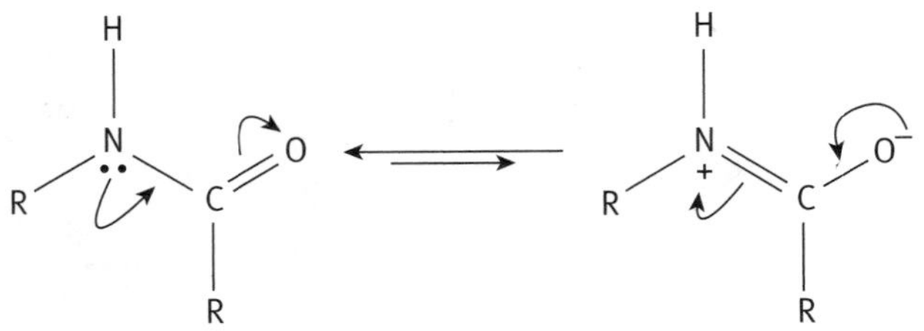

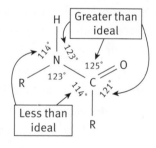

The 3-dimensional structure of the nitrogen in an amide is dictated by its bonding and nonbonding electrons. In an amide, the nitrogen is able to form a resonance hybrid in which it is doubly bound to carbon. Without this resonance form, nitrogen would typically create a trigonal pyramidal geometry, due to repulsion from its lone pair. However, in an amide, those electrons would be shared with the neighboring carbon. This carbon also has a resonant form with a double bond to oxygen, meaning that the nitrogen, oxygen, and carbon must be in plane with one another. This arrangement dictates the trigonal planar geometric form of an amide nitrogen.

2 **Why are the bond angles in a normal amide distorted from the ideal bond angle?**

In a normal amide, the bond shared by carbon and nitrogen is a hybrid of a single and a double bond, meaning the bond contains a relatively large amount of electron density. This electron density repulses the other two atoms bound to nitrogen. As a result of this repulsion, the other substituents of the amide are bent farther away from the nitrogen-carbonyl bond than the ideal bond angle. This bond distortion leads to greater angles on either side of the nitrogen-carbonyl bond, and decreased angle between the other two substituents as a result.

3 **How does the β-lactam structure destabilize the nitrogen-carbonyl bond?**

Takeaways

As the number of lone pairs increases around a central atom, the bond angles in the molecular geometry decrease in magnitude. Multiple bonds can also cause slight distortions in geometry.

Things to Watch Out For

Lone pairs will affect molecular geometry. If lone pairs are not explicitly mentioned for a molecule, you must determine whether lone pairs exist by analyzing the Lewis structure.

A β-lactam contains a four-member ring, a structure you should recognize as being unstable by nature. This instability is caused by the substantial deviation from the ideal bond angle in a 4-membered ring. The typical amide has bond angles approaching the ideal angle of 120°. In a β-lactam ring, the bond angles are heavily distorted, and the carbonyl carbon is held to an angle much smaller than 120° on one side and much larger than 120° on the other. This extreme deviation from ideal molecular geometry destabilizes the nitrogen-carbonyl bond.

 How does penicillin's fused five-membered ring affect the resonance of the amide functional group?

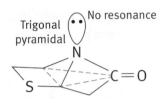

The five-membered ring structure of penicillin is stabilized by maintaining a very specific conformation. The "envelope" shape necessitated by the five-membered ring for structural stability places one corner of the ring out of plane with the rest of the ring, in the "endo" position. In the case of penicillin, nitrogen is kept in the endo position to minimize ring strain. This position forces nitrogen out of plane with both the 5- and 4-membered ring substituents, and functionally creates a trigonal pyramidal geometry for the nitrogen. This structural shape ensures that nitrogen is unable to resonate with the carbon-oxygen double bond of the carbonyl, as it is no longer in plane with the carbon-oxygen double bond.

Related Questions

1. Why is the geometry of aluminum chloride different from that of phosphorus tribromide?

2. Why is the maximum bond angle between adjacent groups in a bent molecule greater than that of an octahedral molecule?

3. Why is the geometry of the sulfate anion the same as that of a sulfuric acid molecule?

High-Yield Problem-Solving Guide questions continue on the next page. ▶ ▶ ▶

E Stoichiometry

During the course of the disproportionation reaction

$$H_2PO_4^- \rightleftharpoons HPO_4^{2-} + H_3PO_4$$

how many grams and molecules of phosphoric acid could be created from 242.5 grams of dihydrogen phosphate? If this amount of phosphoric acid were dissolved in enough water to create one liter of solution, what would the normality of this solution be (with respect to protons)?

1 What is the balanced formula?

The given formula is unbalanced. To balance it, we simply need to double the number of dihydrogen phosphate molecules:

$$2\,H_2PO_4^- \rightleftharpoons HPO_4^{2-} + H_3PO_4$$

2 What are the units when converted to moles?

Stoichiometry is nothing more than an extended application of dimensional analysis. There are four major conversion factors that can be used to convert from the given units to moles (or vice versa):

- **Molar mass, molecular weight**, or **formula weight** (from the Periodic Table): convert from mass (in grams or amu) to moles
- **Avogadro's number** (6.02×10^{23} mol^{-1}): converts from a number of particles to moles
- **Molarity, molality, normality**, or other **measures of concentration**: convert from volume or mass of solution to moles
- **Ideal gases at STP** ($22.4 \frac{L}{mol}$): converts from volume of gas to moles

In this question, we are given a mass (242.5 g), which must be converted to moles. To do this, we need to know the molar mass of dihydrogen phosphate, which can be determined using the Periodic Table. The molar mass is $2 \times 1 \frac{g}{mol} + 1 \times 31 \frac{g}{mol} + 4 \times 16 \frac{g}{mol} = 97 \frac{g}{mol}$.

Now, dividing the given mass by the molar mass gives us the number of moles:

$$\frac{242.5\text{ g}}{97\frac{\text{g}}{\text{mol}}} = 2.5\text{ mol}$$

❸ What is the mol value after accounting for mole ratio?

The next step requires us to use the balanced chemical equation to shift from calculations for dihydrogen phosphate to calculations for phosphoric acid. The mole ratio is simply the stoichiometric coefficient of the desired compound (phosphoric acid) divided by the stoichiometric coefficient of the compound we are given information about (dihydrogen phosphate):

$$2.5\text{ mol H}_2\text{PO}_4^- \times \left[\frac{1\text{ mol H}_3\text{PO}_4}{2\text{ mol H}_2\text{PO}_4^-}\right] = 1.25\text{ mol H}_3\text{PO}_4$$

In other words, reacting 2.5 moles of dihydrogen phosphate will result in 1.25 moles of phosphoric acid.

❹ How can you convert to number of molecules?

We are asked for two different units: mass and number of molecules. Starting with mass, we use the molar mass to convert from moles to grams:

$$1.25\text{ mol H}_3\text{PO}_4 \times \left[\frac{98\text{ g H}_3\text{PO}_4}{1\text{ mol H}_3\text{PO}_4}\right] = 122.5\text{ g H}_3\text{PO}_4$$

This first example points out the utility of setting up all of the conversion factors before multiplying. Consider if Steps 2 through 4 were done in one large multiplication problem:

$$242.5\text{ g H}_2\text{PO}_4^- \times \left[\frac{1\text{ mol H}_2\text{PO}_4^-}{97\text{ g H}_2\text{PO}_4^-}\right]\left[\frac{1\text{ mol H}_3\text{PO}_4}{2\text{ mol H}_2\text{PO}_4^-}\right]\left[\frac{98\text{ g H}_3\text{PO}_4}{1\text{ mol H}_3\text{PO}_4}\right] \approx \frac{240}{2} = 120\text{ g H}_3\text{PO}_4$$

Even with rounding, the answer obtained is still very close to the actual answer.

For the number of molecules, we use Avogadro's number:

$$1.25\text{ mol H}_3\text{PO}_4 \times \left[\frac{6.02 \times 10^{23}\text{ molecules}}{1\text{ mol H}_3\text{PO}_4}\right] = 7.53 \times 10^{23}\text{ molecules}$$

Takeaways

Stoichiometry is an extended version of dimensional analysis that can be answered with a consistent three-step method:

1. Convert from the given units to moles
2. Use the mole ratio
3. Convert from moles to the desired units in the answer choices

5 **What would be the normality of a 1 liter solution containing the H_3PO_4?**

The normality is the molarity of a given molecule multiplied by the number of equivalents of interest given off by one mole of the molecule. For this question, the equivalent of interest is protons. Thus, we simply need to multiply the molarity of the solution by the number of protons that can be given off by phosphoric acid.

The molarity is equal to the moles of solute divided by liters of solution:

$$\frac{1.25 \text{ mol } H_3PO_4}{1 \text{ L}} = 1.25 \ M \ H_3PO_4$$

Because phosphoric acid can give off three protons, the normality is:

$$1.25 \ M \ H_3PO_4 \times 3 = 3.75 \ N \ H_3PO_4$$

Related Questions

1. In the formation of water from diatomic hydrogen and oxygen gases, how many grams and molecules of water are produced when 26 g of hydrogen gas react in the presence of excess oxygen?

2. If butane is combusted in the presence of excess oxygen, resulting in the production of 179.2 L CO_2 at STP, what mass of butane was present at the beginning of the reaction?

3. If 44.8 L of heptane react with 1.204×10^{24} molecules of oxygen in a combustion reaction, how many grams of water are produced?

High-Yield Problem-Solving Guide questions continue on the next page. ▶ ▶ ▶

§ Reaction Rates

In 1940, Sir Christopher Ingold proposed the mechanism for the S_N1 reaction. He used experimentally determined rate laws for a great many S_N1 reactions to validate his theory that the S_N1 reaction follows a multi-step mechanism and that the rate-determining step is carbocation formation. Consider the following S_N1 substitution of tert-butyl bromide carried out in water and the accompanying experimental data about this reaction's rate:

$$H_3C-\underset{\underset{H_3C}{|}}{\overset{\overset{H_3C}{|}}{C}}-Br + CH_3OH + H_2O \longrightarrow H_3C-\underset{\underset{H_3C}{|}}{\overset{\overset{H_3C}{|}}{C}}-OCH_3 + Br^- + H_3O^+$$

	[R₃CBr] (M)	[CH₃OH] (M)	Initial Rate (M/s)
Trial 1	2.0×10^{-1}	4.0×10^{-2}	2.97×10^{-6}
Trial 2	2.0×10^{-1}	8.0×10^{-2}	2.78×10^{-6}
Trial 3	4.0×10^{-1}	8.0×10^{-2}	6.03×10^{-6}

Why do the experimental data prove that carbocation formation is the rate-determining step?

1 What are the three possible elementary reactions and their rate laws?

The general form of the rate law includes a rate constant, k, multiplied by the concentrations of each of the reactants raised to a certain power.

$$\text{rate} = k[R_3CBr]^x[CH_3OH]^y$$

A unimolecular, first-order elementary reaction involves a single unit of one reactant molecule, which dissociates, isomerizes, or otherwise reacts. A bimolecular, and thus second-order, reaction must involve two reacting components, but these could be two different reactants (D and E in the figure below), or two identical reactants (C and C in the figure below).

	Reactants		Products	Rate Law
First Order	A B	$\longrightarrow$	A + B	Rate = $k[AB]^1$
Second Order	C + C	$\longrightarrow$	C C	Rate = $k[C]^2$
	D + E	$\longrightarrow$	D E	Rate = $k[C]^1[C]^1$

2 What are the rate laws for each step of an S_N1 reaction?

Rate laws for each step can be determined using the elementary reaction rules.

Step 1: Carbocation Formation

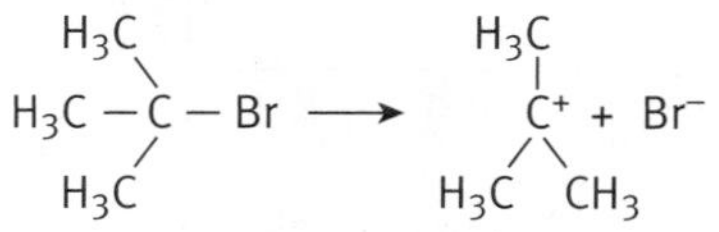

$$H_3C-\overset{\overset{\displaystyle H_3C}{|}}{\underset{\underset{\displaystyle H_3C}{}}{C}}-Br \longrightarrow \overset{\overset{\displaystyle H_3C}{|}}{\underset{\underset{\displaystyle H_3C\quad CH_3}{}}{C^+}} + Br^-$$

Rate = $k[R_3CBr]^1$

Step 2: Nucleophilic Attack

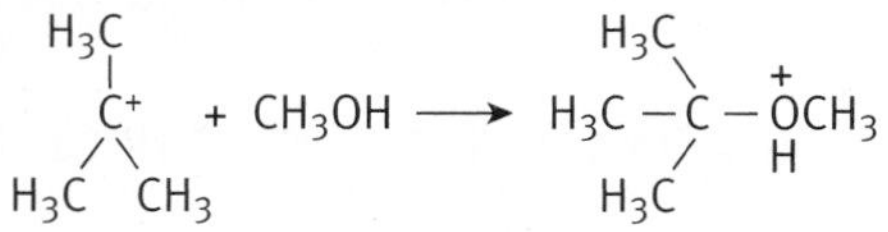

$$\underset{\underset{\displaystyle H_3C\quad CH_3}{}}{\overset{\overset{\displaystyle H_3C}{|}}{C^+}} + CH_3OH \longrightarrow H_3C-\overset{\overset{\displaystyle H_3C}{|}}{\underset{\underset{\displaystyle H_3C}{}}{C}}-\overset{+}{\underset{\underset{\displaystyle H}{}}{O}}CH_3$$

Rate = $k[R_3C^+]^1[CH_3OH]^1$

Step 3: Deprotonation

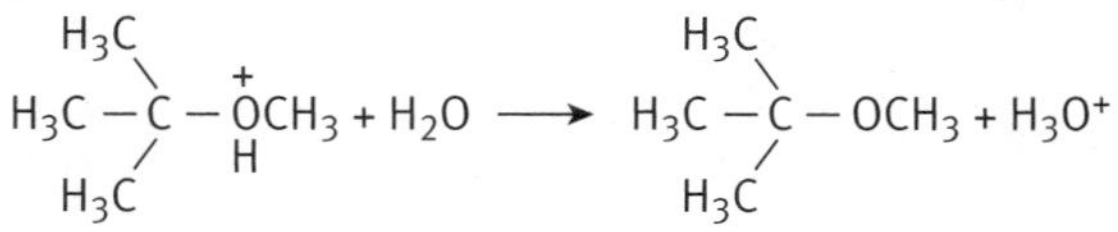

$$H_3C-\overset{\overset{\displaystyle H_3C}{|}}{\underset{\underset{\displaystyle H_3C}{}}{C}}-\overset{+}{\underset{\underset{\displaystyle H}{}}{O}}CH_3 + H_2O \longrightarrow H_3C-\overset{\overset{\displaystyle H_3C}{|}}{\underset{\underset{\displaystyle H_3C}{}}{C}}-OCH_3 + H_3O^+$$

Rate = $k[R3CO^+CH_3]^1[H_2O]^1$

3. Using the data, what is the rate determining step in this reaction?

Begin rate law problems by choosing two trials where only one reactant is changing in concentration. Between trials 1 and 2, $[R_3CBr]$ remains constant and $[CH_3OH]$ doubles in value. When you hold one reactant constant and vary the other, you can set up a proportionality between the initial rate and the concentration of $[CH_3OH]$.

$$\frac{rate_2}{rate_1} = \left[\frac{[CH_3OH]_2}{[CH_3OH]_1}\right]^y$$

$$1 = 2^y$$

$$y = 0$$

Rate was unaffected by change in concentration, giving the rate law an exponent of 0 for $[CH_3OH]$. This allows you to remove $[CH_3OH]$ from the final rate equation. You can repeat this methodology for $[R_3CBr]$ using trials 2 and 3, as $[R_3CBr]$ doubles while $[CH_3OH]$ remains unchanged.

$$\frac{rate_3}{rate_2} = \left[\frac{[R_3CBr]_3}{[R_3CBr]_2}\right]^x$$

$$6.03/2.78 = 2^x$$

$$2 = 2^x$$

$$x = 1$$

You can then plug these values into the overall rate law, yielding:

$$rate = k[R_3CBr]^1[CH_3OH]^0 = k[R_3CBr]^1$$

The data from this experiment demonstrates that, like all S_N1 reactions, the rate is first order.

Takeaways

To determine the order with respect to a particular reactant, use the equation: change in rate = (proportional change in concentration)^x, where x = the order with respect to that reactant.

Related Questions

Things to Watch Out For

Make sure that you select two trials where one reactant's concentration changes but all other concentrations are constant. If there are no two trials where one of the reactants changes, create a new trial using the information you've gleaned about the reaction orders with respect to the *other* reactants.

1. Given the reaction and data in the table below, what is the value of the rate constant k for the nitration of benzene? What are its units?

Trial	$[C_6H_6]$ (M)	$[HNO_2]$ (M)	Initial Rate $\left(\dfrac{M}{s}\right)$
1	1.01×10^{-3}	2.00×10^{-2}	5.96×10^{-6}
2	4.05×10^{-3}	2.00×10^{-2}	5.96×10^{-6}
3	3.02×10^{-3}	6.01×10^{-2}	5.40×10^{-5}

2. Given the data below, determine the rate law for the reaction of pyridine with methyl iodide. Find the rate constant k for this reaction and its units.

Trial	$[C_5H_5N]$ (M)	$[MeI]$ (M)	Initial Rate $\left(\dfrac{M}{s}\right)$
1	1.00×10^{-4}	1.00×10^{-4}	7.50×10^{-7}
2	2.00×10^{-4}	2.00×10^{-4}	3.00×10^{-6}
3	2.00×10^{-4}	4.00×10^{-4}	6.00×10^{-6}

3. Cerium(IV) is a common inorganic oxidant. Determine the rate law for the following reaction and compute the value of the rate constant k along with its units:

$$Ce^{4+} + Fe^{2+} \rightarrow Ce^{3+} + Fe^{3+}$$

Trial	$[Ce^{4+}]$ (M)	$[Fe^{2+}]$ (M)	Initial Rate $\left(\dfrac{M}{s}\right)$
1	1.10×10^{-5}	1.80×10^{-5}	2.00×10^{-7}
2	1.10×10^{-5}	2.80×10^{-5}	3.10×10^{-7}
3	3.40×10^{-5}	2.80×10^{-5}	9.50×10^{-7}

⊟ Reaction Energy Profiles

When chalcone (**A**) is subjected to reductive conditions with sodium borohydride, two products can result. The two products are the so-called 1,2-reduction product (**B**), in which the carbonyl is reduced, and the 1,4-reduction product (**C**), in which the conjugated alkene is reduced.

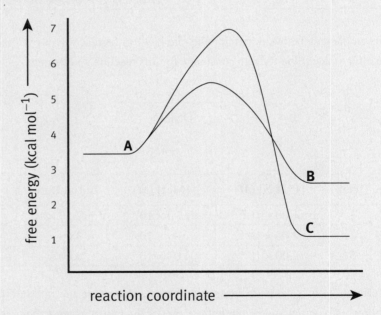

The reaction profiles leading to each reduction product at 298 K are both shown in the plot below:

Based on the plot above, answer the following questions:

1. Which product is more thermodynamically stable? Which one forms faster?

2. Assume that **A** is in equilibrium with **C**. What will the ratio of **C** to **A** be at equilibrium?

3. How could the rate of the reaction of **A** to **C** be made closer to the rate of the reaction of **A** to **B**?

4. Which product would be favored if **A** were subjected to high temperatures for a long time? If **A** were subjected to low temperatures for only a brief period of time? Explain why for each situation.

$$\left(\text{Note: R} = 1.99 \, \frac{\text{cal}}{\text{mol} \cdot \text{K}} \right)$$

1 What energy differences do you notice in the graph?

Notice that the energy of **C** is lower than that of **B**. Therefore, it is the more thermodynamically stable product.

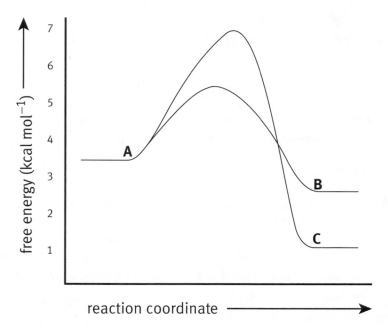

The rate of formation of each product is determined by the difference in energy between the starting material **A** and the top of the hump leading to each product. Because this difference is lower for the formation of **B**, it forms faster.

2 How does difference in energy relate to ratio of products to reactants at equilibrium?

The relevant equation relating the change in free energy to the concentrations of reactants and products at equilibrium is:

$$\Delta G^\circ = -RT \ln K_{eq}$$

We can rearrange it to solve for K_{eq}:

$$\ln K_{eq} = \frac{-\Delta G^\circ}{RT}$$

$$K_{eq} = e^{\frac{-\Delta G^\circ}{RT}}$$

Note from the diagram that $\Delta G° \approx 1000 - 3500 = -2500 \frac{cal}{mol}$. We are told that $R = 1.99 \frac{cal}{mol \cdot K}$ and that $T = 298$ K. Plugging into the equation, we get:

$$K_{eq} \approx e^{\frac{2500 \frac{cal}{mol}}{(2 \frac{cal}{mol \cdot K})(300 \text{ K})}} = e^{\frac{25}{6}}$$

Let's assume that $25 \div 6$ is about equal to 4, and that e (2.718…) is about equal to 3. With these simplifications, we can estimate:

$$K_{eq} = \frac{[\mathbf{C}]}{[\mathbf{A}]} \approx 3^4 = 81 \text{ (actual} = 68)$$

Note that $\Delta G° < 0$ gives more product than reactant, as one would expect for a spontaneous reaction containing only one reactant and one product.

3 How can activation energy be lowered for the reaction A → C?

The most promising way to bring the rate of **A** to **C** closer to that of **A** to **B** is to lower the activation energy of the **A** to **C** reaction. This could be accomplished by adding a catalyst for the **A** to **C** reaction. Remember: a catalyst speeds up the reaction but is not consumed in the process.

4 How would temperature affect the reactions?

At high temperatures for a long time, **A** has sufficient energy to keep reacting over and over again to form **B** and **C**, which can then react back to reform **A**. Over time, the lowest energy (thermodynamic) product, **C**, would predominate because it is the most stable.

Between the two products, **B** will form much faster than **C** because its activation energy (the height of the hump) is lower. At low temperatures, there may not be sufficient energy for **B** to revert back to **A**, or for **A** to convert to **C**. Therefore, the faster-forming (kinetic) product, **B**, will predominate.

Takeaways

The goal of a reaction profile is to give information about energy *differences*. Make sure to identify the important differences and their significances, as in this problem.

Things to Watch Out For

Be careful to take note of the units of energy on the *y*-axis if there are any necessary computations. The *y*-axis could be given in joules, calories, or Calories (kilocalories).

Related Questions

1. What would be the ratio of **B** to **A** at equilibrium?

2. If a catalyst were added to the reaction of **A** going to **C**, as in Step 3 earlier, would the energies of **A** and **C** be changed as a result? Why or why not?

3. There are actually intermediates involved in the reactions producing both **B** and **C**, which are shown below. Sketch how each reaction profile would look, including the involvement of these intermediates. Be sure to indicate which intermediate is relatively more stable.

Key Concepts

E Hess's Law

The Nutrition Labeling and Education Act (NLEA) gives the Food and Drug Administration (FDA) the authority to require most foods regulated by the FDA to bear nutrition labeling. Among other information, calories must appear in bold print on the labels. To determine the number of calories of a food, scientists burn it in a bomb calorimeter. A measured food sample is inserted in its inner chamber, after which the inner chamber is filled with oxygen and sealed. The outer chamber contains a certain amount of cold water. An electric spark ignites the inner chamber, and the resulting change in water temperature is measured to determine the food's heat of combustion.

Substance	ΔH_f° (kJ/mol)
CO_2	−393.5
H_2O	−241.8
$C_5H_{10}O_5$ (xylose)	−1057.8
$C_6H_{12}O_6$ (glucose)	−1274.5

A student attempts to experimentally determine the heat released by the combustion of 10 grams of glucose ($C_6H_{12}O_6$) and 10 grams of xylose ($C_5H_{10}O_5$) using a calorimeter. Provided the experiment goes flawlessly, how will the two measured values compare?

1 How can a balanced reaction for the combustion of any carbohydrate be written?

Because sugars have a 1:2:1 ratio of carbon, hydrogen, and oxygen, respectively, their empirical formula is CH_2O. Therefore, they can be written as $C_nH_{2n}O_n$ (where n is an integer ≥ 3). The unbalanced reaction below is typical of all carbohydrate combustion reactions that involve oxygen gas:

$$C_nH_{2n}O_n + O_2 \rightarrow CO_2 + H_2O$$

Begin by balancing the carbon atoms on the right side (n CO_2), and then balance the hydrogen atoms on the right side (n H_2O). Finally, balance the oxygen atoms on the left side (n O_2):

$$C_nH_{2n}O_n + n\,O_2 \rightarrow n\,CO_2 + n\,H_2O$$

This procedure can be applied to determine the balanced reaction of xylose and glucose combustion reactions:

(xylose) $C_5H_{10}O_5 + 5\,O_2 \rightarrow 5\,CO_2 + 5\,H_2O$

(glucose) $C_6H_{12}O_6 + 6\,O_2 \rightarrow 6\,CO_2 + 6\,H_2O$

2 How does the heat of formation of glucose compare to that of xylose?

The standard heats of formation of xylose and glucose are −1057.8 kJ/mol and −1274.5 kJ/mol, respectively. In order to determine how they compare, it's best to approximate the values to more manageable numbers, truncating down to −1000 and −1200, respectively.

$$\frac{\Delta H_f^\circ \text{ xylose}}{\Delta H_f^\circ \text{ glucose}} = \frac{-1057.8 \text{ kJ/mol}}{-1274.5 \text{ kJ/mol}} \approx \frac{-1000 \text{ kJ/mol}}{-1200 \text{ kJ/mol}} = \frac{5}{6}$$

3 How can the ΔH_f° of oxygen be determined?

The standard heat of formation, ΔH_f°, is the amount of thermal energy required to synthesize a compound from its elements in their standard states (the phase in which the element exists at room temperature, 298 K). The heat of reaction, ΔH_{rxn}°, is the amount of thermal energy required to convert the reactants to the products of a reaction. Hess's law states that the difference between the heats of formation of the products and the reactants is equal to the heat of the reaction.

$$\Delta H_{rxn}^\circ = \text{sum of } \Delta H_{f,\text{products}}^\circ - \text{sum of } \Delta H_{f,\text{reactants}}^\circ$$

Hess's law can be used to determine the standard heat of combustion of glucose.

$$\Delta H_{rxn}^\circ = \text{sum of } \Delta H_{f,\text{products}}^\circ - \text{sum of } \Delta H_{f,\text{reactants}}^\circ$$

$$\Delta H_{comb}^\circ = \left(6 \times \Delta H_{f\,H_2O}^\circ + 6 \times \Delta H_{f\,CO_2}^\circ\right) - \left(\Delta H_{f\,\text{glucose}}^\circ + 6 \times \Delta H_{f\,O_2}^\circ\right)$$

$$\Delta H_{comb}^\circ = \left(6 \times -241.8 \text{ kJ/mol} + 6 \times -393.5 \text{ kJ/mol}\right) - \left(-1274.5 \text{ kJ/mol} + 6 \times \Delta H_{f\,O_2}^\circ\right)$$

$$\Delta H_{comb}^\circ = -2805 \text{ kJ/mol} - 6 \times \Delta H_{f\,O_2}^\circ$$

Oxygen is already in its standard state (a diatomic gas). Accordingly, its standard heat of formation is zero. Plugging this in the calculations above, the standard heat of combustion of glucose is −2805 kJ/mol.

4 How can the standard heat of combustion of xylose be determined with limited calculations?

To calculate the heat of combustion for xylose, we can simply adapt the second line of calculations of the previous prompt by making use of the 5/6th ratio of coefficients:

$$\Delta H_{comb}^\circ = \left(5 \times \Delta H_{f\,H_2O}^\circ + 5 \times \Delta H_{f\,CO_2}^\circ\right) - \left(\Delta H_{f\,\text{xylose}}^\circ + 5 \times \Delta H_{f\,O_2}^\circ\right)$$

As previously determined, the heat of formation of xylose is 5/6 that of glucose. Consequently, xylose's ΔH°_{comb} will necessarily be approximately 5/6 that of glucose, -2337.5 kJ/mol.

5 How can we compare the ΔH_{comb} of 10 g of each sugar without calculating molar mass?

Standard conditions assumes one mole of reactants. In order to determine the heat released by 10 grams of carbohydrate from its ΔH°_{comb}, we need to use molar mass. We can avoid using extensive calculations by making use of ratios. Xylose's molecular formula is $C_5H_{10}O_5$, while glucose's is $C_6H_{12}O_6$. Glucose has 6/5 more of each atom present in xylose, making its molar mass 6/5 that of xylose.

6 What is the ratio of the combustion of ten grams of xylose to that of glucose?

We multiply the standard heats of combustion of each sugar by the inverses of their molar masses to determine the heat 1 gram of each of these sugars releases. Multiplying the product by 10 gives the heat released by the combustion of 10 grams of these sugars.

Heat of combustion of 10 g of glucose: Heat of combustion of 10 g of xylose:

$$\Delta H_{\text{rxn, 10 grams}} = \frac{\Delta H^{\circ}_{comb}}{1\ \text{mol}} \times \frac{1\ \text{mol}}{MM} \times 10\ \text{g} \qquad \Delta H_{\text{rxn, 10 grams}} = \frac{\Delta H^{\circ}_{comb}}{1\ \text{mol}} \times \frac{1\ \text{mol}}{MM} \times 10\ \text{g}$$

The ratio of the sugars' standard heats of combustion is 5:6 (xylose:glucose). Substituting those values in, we find that the heat released by the combustion of 10 grams of these sugars is equal.

Heat of combustion of 10 g of glucose: Heat of combustion of 10 g of xylose:

$$\Delta H_{\text{rxn, 10 grams}} = \frac{6}{1\ \text{mol}} \times \frac{1\ \text{mol}}{6} \times 10\ \text{g} \quad = \quad \Delta H_{\text{rxn, 10 grams}} = \frac{5}{1\ \text{mol}} \times \frac{1\ \text{mol}}{5} \times 10\ \text{g}$$

Therefore, Hess's law explains why carbohydrates have the same caloric properties (4 cal/g), irrespective of the carbohydrate consumed. You can use similar math to prove that all lipids produce 9 cal/g, and all proteins 4 cal/g. Across all three classes

Takeaways

Always determine the balanced equation for the reaction before you begin to apply Hess's law. Recall that enthalpy is a state function, so regardless of the path one takes to get from the reactants to the products, the change in enthalpy will be the same.

Things to Watch Out For

At least one of the wrong answer choices for thermochemistry questions will be the result of carelessness with signs. Carefully organized scratchwork will help you avoid this problem, but perhaps more important is the ability to round judiciously. Only experience (practice) will breed such wisdom.

of major biomolecule, bigger molecules have higher standard heats of combustion, but proportionally less of them "fit" in one gram. So, while a molecule of glucose may have a larger mass than a molecule of xylose, 10 grams of glucose produces *exactly* as much energy within your body as does 10 grams of xylose!

Related Questions

1. Given the ΔH_f° of carbon dioxide and water, what other piece(s) of information would you need to have to calculate the ΔH_{comb}° of ethane?

2. If the ΔH_f° of acetylene is 226.6 $\frac{kJ}{mol}$, what is the ΔH_{comb}° of acetylene?

3. If the ΔH_f° of NaBr (s) is −359.9 $\frac{kJ}{mol}$, what is the sum of each ΔH_{rxn}° of the following series of reactions?

$$Na\ (s) \rightarrow Na\ (g) \rightarrow Na^+\ (g)$$

$$\frac{1}{2} Br_2\ (g) \rightarrow Br\ (g) \rightarrow Br^-\ (g)$$

$$Na^+\ (g) + Br^-\ (g) \rightarrow NaBr\ (s)$$

Ideal Gases

The ideal gas law is used to characterize the behavior of theoretical gases that have no size or intermolecular effects. In the real world, this law is often confounded by additional effects, leading to the creation of the Van der Waals equation, which adjusts for some of the additional effects observed in real-world gases. Nitrous oxide (N_2O) is an anesthetic gas typically stored in small, pressurized canisters and administered via face mask. Even though the ideal gas law does allow for approximation of moles of compound (n), the ideal gas law cannot be used to correctly calculate the concentration of nitrous oxide gas. Why is it that nitrous oxide's behavior cannot be correctly characterized by the ideal gas law?

1 How are *P*, *V*, and *T* related to *n*?

The ideal gas law is defined as:

$$PV = nRT$$

In the ideal gas law, P = pressure, V = volume, n = number of moles of gas, R = the Rydberg constant, and T = temperature. When considering the impact of variation of P, V, or T on the value of n, the following mathematical relationships can be established:

$$P \propto n$$

$$V \propto n$$

$$1/T \propto n$$

Pressure and volume both vary directly with number of moles of gas, but temperature varies inversely. When using these mathematical relationships, you should keep "real world" considerations in mind. For example, pressure and moles of gas are directly related. But while increasing the number of moles of gas in a fixed-volume container will increase pressure, increasing pressure does *not*, in the real world, increase the number of moles of gas!

2 How would *P*, *V*, and *T* affect each other in an ideal gas, using Boyle's and Charles's laws?

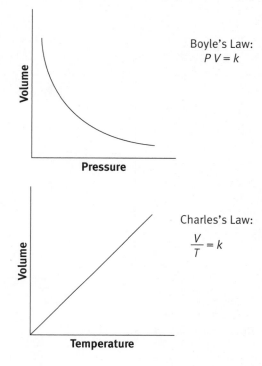

Boyle's Law:
$$PV = k$$

Charles's Law:
$$\frac{V}{T} = k$$

Boyle's law demonstrates mathematically that pressure and volume vary inversely with one another. Charles's law shows a direct relationship between volume and temperature. These relationships allow for changes in pressure, volume, and temperature to occur without changing the value of constants R and *n*.

3 **How would nitrous oxide deviate from the behavior of a more ideal gas (such as helium)?**

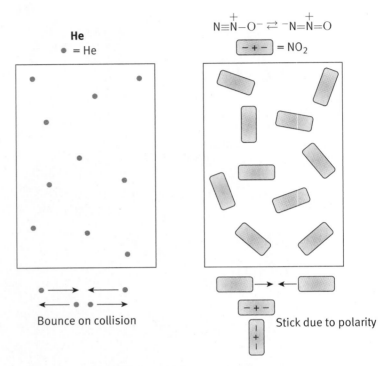

*Note: molecules not drawn to scale

The molecules of nitrous oxide take up substantially more volume than the helium molecules. The ideal gas law assumes negligible size of the molecule, so that all volume occupied by the gas can be dictated by the ideal equation. Nitrous oxide, due to its size, would deviate substantially from the predicted volume due to molecule size.

Nitrous oxide possesses a distinctive dipole within the molecule due to its resonant structure and comparative electronegativity. Upon collision, helium and other near-ideal gases would be expected to bounce perfectly, as there are no potential sources of strong intermolecular forces. The dipole of nitrous oxide would lead to increased stability with alignment of positive charges adjacent to nearby negative charges. These so-called "sticky interactions" between molecules lead to further deviation from ideal behavior, as nitrous oxide will stay near other nitrous oxide molecules in T–shape alignment, lowering the pressure exerted on the walls of the container.

How might N₂O behave in a small, pressurized canister?

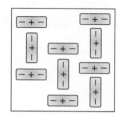

Nitrous oxide displays intermolecular forces even when in a large quantity of space. When space is reduced and temperature is dropped, these intermolecular forces and "sticking" effects will increase. Functionally, if stored at cool enough temperatures and a small enough volume, nitrous oxide would behave as a liquid, rather than a gas. Converting nitrous oxide to a liquid for storage would allow for new types of measurements to be taken based on nitrous oxide's behavior as a liquid, avoiding the issue of problems calculating its dosage as a gas.

Related Questions

1. A 22.4 liter vessel contains nitrogen gas at STP. What mass of nitrogen is in the vessel? $\left(\text{Note: } R = 8.314 \dfrac{J}{mol \cdot K}\right)$

2. Humans breathe out a tidal volume of about 0.5 liters of gas at 1.0 atm and 37°C. How many moles of gas are present? (Note: Assume that only carbon dioxide is present in the exhaled gas and that $R = 0.0821 \dfrac{L \cdot atm}{mol \cdot K}$)

3. What is the volume of 2 mol of chlorine gas pressurized to 4000 mmHg at 700°C? $\left(\text{Note: } R = 62.4 \dfrac{L \cdot mmHg}{mol \cdot K}\right)$

S Solution Equilibria

The molar solubility of iron (III) hydroxide in pure water at 25°C is 9.94×10^{-10} M. How would the substance's molar solubility change if placed in aqueous solution with a pH of 10.0 at 25°C?

1 **Why is the solubility product, K_{sp}, of this dissolution reaction less than the molar solubility of iron (III) hydroxide?**

For this reaction, the generic dissociation reaction is:

$$Fe(OH)_3\,(s) \rightleftharpoons Fe^{3+} + 3OH^-$$

This reaction can be envisioned as a solid dissociating into four daughter particles:

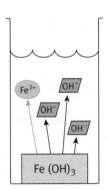

Utilizing this visual of the reaction, you can conclude that four moles of ions are dissolved into the solution per mole of solid dissolved. This reaction can also be considered quantitatively. K_{sp} is an equilibrium constant, just like any other K, and is given the special name solubility product because it is related to the dissolution of salt in a solution. The expression for any K is:

$$K_{eq} = \frac{[\text{Products}]}{[\text{Reactants}]}$$

in which the concentration of each product is raised to its stoichiometric coefficient over the concentration of each reactant raised to its stoichiometric coefficient. K-values are constant for a given temperature. Because the denominator of K_{sp} is a solid, and solids are not included in equilibrium constants, K_{sp} will never include a formal denominator. This exclusion makes K_{sp} a special case among equilibrium constants. For iron (III) hydroxide, the K_{sp} expression is:

$$K_{sp} = [Fe^{3+}][OH^-]^3$$

For every x moles of iron (III) hydroxide to dissolve, x moles of iron (III) and $3x$ moles of hydroxide ions will be dissolved into solution. This proportion can also be seen in the visual representation above. For example, if x were 2 moles, then 2 moles of iron (III) cation and 6 moles of hydroxide anion would be produced when iron (III) hydroxide dissolves. It should be noted that in practice on the MCAT the value of x, or molar solubility, of a compound used for K_{sp} calculations will always be substantially below the value of 1. K_{sp} will always be less than 1 because K_{sp} calculations are utilized specifically for compounds with low solubility on Test Day. The K_{sp} expression can be rewritten:

$$K_{sp} = [x][3x]^3 = 27x^4$$

The molar solubility, x, is given in the question stem as 9.94×10^{-10}. Plugging in, the value of K_{sp} is:

$$27 \times (9.94 \times 10^{-10})^4 \approx 27 \times (10^{-9})^4 = 27 \times 10^{-36} = 2.7 \times 10^{-35}$$

If you consider your final output value for K_{sp} as compared to molar solubility in the context of the question stem, you can see that K_{sp} is indeed a smaller value than molar solubility x. Knowing that every problem of this type will contain a value of x substantially below 1, and that by definition K_{sp} will always involve at least two product ions in the K_{sp} equation, by nature the value of x will be greater than K_{sp}, and both will be substantially below 1. This relationship is not only a persistent rule, but is also one that can be used to great effect for fast triaging on the MCAT.

Takeaways

The value of K_{sp} does not change when a common ion is present; it is a constant that is dependent on temperature only. The molar solubility of the salt, however, does change if a common ion is present.

2 Why does an increase in pH substantially decrease the molar solubility of iron (III) hydroxide?

When pH is increased above the standard value of 7, we can conclude there is now an increase of hydroxide ions in solution via the acid constant equation:

$$pH + pOH = 14$$

Things to Watch Out For

Be careful when applying Le Châtelier's principle in cases of precipitation and solvation. For a solution at equilibrium (that is, a saturated solution), adding more solid will not shift the equilibrium to the right. More solid does not dissociate to raise the ion concentrations; it just precipitates to the bottom.

Thus, the concentration of [OH⁻] present in solution for use in the K_{sp} expression would increase substantially as well:

$$K_{sp} = [Fe^{3+}][OH^-]^3$$

Because K_{sp} is constant, an increase in [OH⁻] means [Fe³⁺] must decrease. The effect is that less iron (III) hydroxide can dissociate. You can also consider this shift qualitatively, via Le Châtelier's principle: as product concentration is increased, reactant is favored. Ultimately, in any solubility product calculation, the introduction of any ion common to the products will cause a decrease in molar solubility. This change in solubility is referred to as the common ion effect.

Related Questions

1. Given a substance's K_{sp}, how could you solve for its molar solubility in pure water? What if a common ion were also present in solution?

2. Given a table listing substances and their solubility constants, how could you determine which substance was most soluble in pure water?

3. Given that the sulfate ion can react with acid to form hydrogen sulfate, how would the molar solubility of sulfate salts be affected by varying a solution's pH?

High-Yield Problem-Solving Guide questions continue on the next page. ▶ ▶ ▶

S Acids and Bases

Of the hydrogen halide series of acids (HF, HCl, HBr, and HI) only HF is classified as a "weak" acid; the other hydrogen halide acids—for example, HCl—are classified as "strong." Yet, HF is considerably more dangerous to human health than is HCl. According to the US Centers for Disease Control, "even small splashes of concentrated HF may be fatal." A patient exposed to concentrated HF will develop severe tissue damage at the exposure site, but this damage will develop slowly over a period of several hours. HF may even penetrate deeply enough to cause irreversible bone damage or, in fatal cases of exposure, HF will attack major organs, causing systemic organ failure. By contrast, exposure to HCl, even concentrated HCl, is rarely fatal. Concentrated HCl may cause immediate, often severe, chemical burns. However, these burns are usually superficial and may be immediately treated with running water, followed by a standard burn protocol. Finally, seemingly paradoxically: It is the very fact that HF is a "weak" acid that makes it so physiologically damaging. Resolve this apparent paradox using physiology and Brønsted–Lowry acid-base chemistry.

1   What explains the difference in HF- and HCl-mediated injuries?

Hydrofluoric acid has a K_a less than 1, while hydrochloric acid has a K_a greater than 1. K_a indicates how much an acid dissociates. The higher its K_a value, the stronger the acid is. HF's small K_a indicates that it dissociates fairly poorly. Since protons are the damage-inducing agents in this system, HCl causes immediate damage by rapidly and completely dissociating in solution, while HF takes longer to release its protons into solution. HF continuously dissociates because as it releases protons, the protons are used up in other physiological reactions. Le Châtelier's principle predicts that removing reactants (in this case, protons) causes a shift in equilibrium. Over a period of several hours, the HF equilibrium will continue to shift to the right as hydrofluoric acid's protons are pulled off.

$$HF + H_2O \rightleftharpoons F^- + H_3O^+$$
$$K_a \approx 10^{-5}$$

$$HCl + H_2O \rightleftharpoons Cl^- + H_3O^+$$
$$K_a \approx 10^{6}$$

2 How do the definitions of "weak" and "strong" acids correspond to the *K*-values of water and hydronium?

An acid whose K_a is greater than 1 is a strong acid. An acid whose K_a is less than 1 is a weak acid. This distinction is based on the observation that the hydronium ion has a K_a of 1. For an acid to have a K_a greater than 1, that acid must be more reactive than the hydronium ion. And again, these acids are considered "strong."

$$H_3O^+ + H_2O \rightleftharpoons H_2O + H_3O^+$$
$$K_a = [H_3O^+]/[H_3O^+] = 1$$

By contrast, a molecule of water can act as an "acid" by donating a proton to another molecule of water. The *K*-value for this reaction is $K_w = 10^{-14}$. Put another way: Water is extremely unlikely to auto-ionize and donate a proton to another water molecule, though this does happen. Scientists therefore take this value, the K_w value, as the other extreme of acidity—any proton donor with a K_a less than K_w is considered "inert" as an acid. Ultimately, the *K*-values for hydronium and for water define acidity: If an acid's K_a is greater than the *K*-value for hydronium, that acid is considered "strong." If an acid's K_a is less than the *K*-value for hydronium, but more than the K_w value for water, that acid is considered "weak." And if a molecule's K_a is less than K_w, that molecule is considered "inert as an acid."

$$H_2O + H_2O \rightleftharpoons OH^- + H_3O^+$$
$$K_w = [H_3O^+][OH^-] = 1 \times 10^{-14}$$

3 According to Brønsted–Lowry chemistry, what makes a base "strong" or "weak"?

Part of what accounts for the physiological danger of HF is the reactivity of HF's "weak" conjugate base. Similarly to hydronium, the hydroxide's K_b value of 1 marks the cutoff between a weak and strong base. If a base's K_b is lower than 1, it is considered a weak base, while a K_b higher than 1 is indicative of a strong base.

$$OH^- + H_2O \rightleftharpoons OH^- + H_2O$$
$$K_b = [OH^-]/[OH^-] = 1$$

4   Why is the deprotonated form of HF dangerously reactive while Cl⁻ is not?

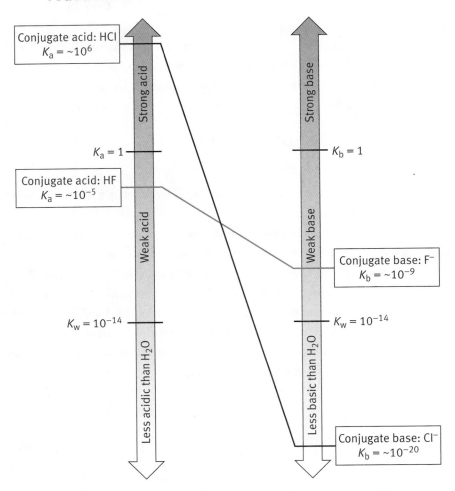

We can use the relationship between the K_a of an acid and the K_b of its conjugate base to explain the relative danger of HF. HF is a weak acid. Calculating its conjugate base's K_b indicates that F^- is a weak base ($K_b = 10^{-9}$). A base that is weak is not unreactive. Conversely, HCl has a K_a value of 10^6, meaning that the K_b of Cl⁻ is extremely low ($K_b = 10^{-20}$). This vanishingly small K_b means Cl⁻ is an inert base, even less basic than H_2O. Thus, in the body, F^- is able to disrupt physiological functions, while Cl⁻ is not. As an acid gets more reactive, its conjugate base becomes proportionally less reactive.

HF vs. F^-

$$K_a \times K_b = K_w$$
$$10^{-5} \times K_b = 10^{-14}$$
$$10^{-14}/10^{-5} = K_b = 10^{-9}$$

HCl vs. Cl⁻

$$K_a \times K_b = K_w$$
$$10^6 \times K_b = 10^{-14}$$
$$10^{-14}/10^6 = K_b = 10^{-20}$$

5 Why can HF penetrate physiological membranes and cause deep tissue damage?

Time to put all our work into practice! As soon as HCl is put in solution, it dissolves, meaning it is only present in the form of its ions, H^+ and Cl^-. H^+ is physiologically damaging, however charged ions like H^+ cannot go through cellular membranes. So, the fact that HCl is a strong acid means it is superficially damaging, but it does not penetrate deeply. By contrast, HF dissociates extremely slowly, and that is precisely because it is a "weak" acid. As a result, in solution, HF remains HF; it does not dissociate into charged ions. Lacking charge, HF is able to penetrate physiological membranes. So, HF can penetrate into your cells, your organs, your bones, and your bloodstream. And once it has invaded a target cell, *then* HF dissociates into its ions, H^+ and F^-, both of which are capable of causing physiological damage!

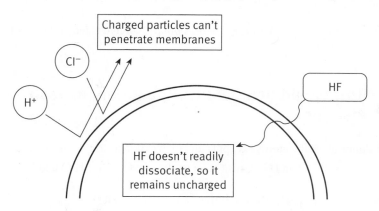

Takeaways

You are likely to see six main types of equilibrium constants on Test Day: K_{eq}, K_{sp}, K_f, K_a, K_b, and K_w. The five latter are all a subtype of the former in that all K values represent the system at equilibrium. K_a specifically indicates the strength of an acid: the more it dissociates in solution, the higher its K_a value.

Things to Watch Out For

Be careful when inferring the properties of conjugate acids or bases. The conjugate acid of a weak base is not necessarily strong. In addition, a compound being weak doesn't mean it is inert.

Related Questions

1. What can be assumed about the conjugate acid of a strong base?

2. What can be assumed about the conjugate acid of a weak base?

3. If the pK_a of acetic acid is 4.76, then what is the pK_b of acetate?

Key Concepts

Balancing Oxidation–Reduction
E Reactions

Balance the following reaction, which takes place in basic solution:

$$ZrO(OH)_2 \, (s) + SO_3^{2-} \, (aq) \rightleftharpoons Zr \, (s) + SO_4^{2-} \, (aq)$$

1 What are the two half-reactions?

$$ZrO(OH)_2 \rightleftharpoons Zr$$
$$SO_3^{2-} \rightleftharpoons SO_4^{2-}$$

Break the reactions up by looking at atoms *other* than hydrogen and oxygen.

2 How would you balance the atoms of each half reaction?

First, balance all of the atoms except H and O. Next, in a basic solution, use OH^- and H_2O to balance the O and H atoms (in an acidic solution, use H_2O and H^+):

$$H_2O + ZrO(OH)_2 \rightleftharpoons Zr + 4\,OH^-$$
$$2\,OH^- + SO_3^{2-} \rightleftharpoons SO_4^{2-} + H_2O$$

3 How would you balance the charges of each half reaction?

Add electrons as necessary to one side of each reaction so that the charges are equal on both sides. The top equation has a total charge of –4 on the right from the 4 hydroxide ions, so 4 electrons need to be added to the left side of the equation. In the bottom equation, there is a total charge of –4 on the left (–2 from the two hydroxide ions and –2 from the sulfite anion) and –2 on the right (from the sulfate anion):

$$4\,e^- + H_2O + ZrO(OH)_2 \rightleftharpoons Zr + 4\,OH^-$$
$$2\,OH^- + SO_3^{2-} \rightleftharpoons SO_4^{2-} + H_2O + 2\,e^-$$

During this step, be careful to account for all charges, including the charges contributed by molecules other than H^+ and OH^-.

4 How would you balance the electrons of each half reaction?

Here, the lowest common multiple between the four electrons in the top reaction and the two in the bottom is four electrons, so we must multiply everything in the bottom reaction by two:

$$4\,e^- + H_2O + ZrO(OH)_2 \rightleftharpoons Zr + 4\,OH^-$$
$$4\,OH^- + 2\,SO_3^{2-} \rightleftharpoons 2\,SO_4^{2-} + 2\,H_2O + 4\,e^-$$

5 What is the combined reaction?

$$ZrO(OH)_2 + 2\,SO_3^{2-} \rightleftharpoons Zr + 2\,SO_4^{2-} + H_2O$$

6 How can you confirm that the reaction is properly balanced?

There is a −2 net charge on each side of the reaction equation, and the atoms are stoichiometrically balanced. This last step is extremely important. If mass and charge aren't balanced, then we have made an error in one of the previous steps.

Takeaways

Don't fall into the trap of simply balancing mass in these reactions. If oxidation and reduction are occurring, one must go through this procedure to balance the reaction accurately.

Things to Watch Out For

These kinds of problems can be extremely tedious. Questions that ask you to balance complicated oxidation–reduction reactions are best saved until the end of the section if you have time left over.

Related Questions

1. Which atom is being oxidized in the original equation? Which is being reduced? Identify the oxidizing and reducing agents.

2. A disproportionation reaction is one in which the same species is both oxidized and reduced during the course of the reaction. One such reaction is shown below. Balance the reaction, assuming that it takes place in acidic solution:

$$PbSO_4\ (s) \rightarrow Pb\ (s) + PbO_2\ (s) + SO_4^{2-}\ (aq)$$

3. Dentists may use zinc amalgams to make temporary crowns for their patients. It is absolutely vital that they keep the zinc amalgam dry. The reaction of zinc metal with water is shown below:

$$Zn\ (s) + H_2O\ (l) \rightarrow Zn^{2+}\ (aq) + H_2\ (g)$$

Balance this reaction, assuming that it takes place in basic solution. Why is it so important to keep the amalgam dry?

Key Concepts

General Chemistry Chapter 12
Oxidation–reduction reactions
Electrochemical cells
Stoichiometry
Electromotive force
$\Delta G = -nFE^\circ_{cell}$

$\boxed{\text{S}}$ Electrochemical Cells

The mitochondrion's electron transport chain (ETC) is a series of compounds that transfer electrons from electron donors to electron acceptors via redox reactions. Thus, this organelle can be thought of as an electrochemical cell. The ETC's first reaction is the oxidation of NADH, during which complex I catalyzes the transfer of electrons from NADH to CoQ10. The reduced ubiquinone (CoQ10) subsequently transports these electrons to the cytochrome b-c complex (part of complex III). Below are standard potentials for NADH, cytochrome b and cytochrome c.

$NADH \rightarrow NAD^+ + 2e^- + H^+$ $\qquad E^\circ = +0.320 \text{ V}$
$2 \text{ cytochrome } b_{(red)} \rightarrow 2 \text{ cytochrome } b_{(ox)} + 2e^-$ $\qquad E^\circ = -0.070 \text{ V}$
$2 \text{ cytochrome } c_{(ox)} + 2e^- \rightarrow 2 \text{ cytochrome } c_{(red)}$ $\qquad E^\circ = +0.254 \text{ V}$

What is the chronological order in which the three compounds above are oxidized in the ETC?

1 Which half-reaction is given a different type of standard potential than the others?

There are two types of standard potentials: reduction and oxidation potentials. Considering the three half-reactions above, only one of them is an oxidation reaction (that of cytochrome b). Accordingly, this reaction is associated with cytochrome b's oxidation potential. The reduction potentials are provided for both NADH and cytochrome c.

2 Compare the reduction potentials of the following electron acceptors to cytochrome c's.

The following electron acceptors in the ETC are shown below in chronological order (1 → 2 → 3). Oxygen acts as the final electron acceptor in the ETC. Therefore, oxygen must be the most likely compound to accept electrons. In other words, oxygen must have the highest reduction potential. The same logic can be applied when looking at cytochrome a3, the second-to-last electron acceptor: this cytochrome variant should have the second-highest reduction potential when considering all the electron acceptors of the ETC.

1. $2 \text{ cytochrome } c_{(ox)} + 2e^- \rightarrow 2 \text{ cytochrome } c_{(red)}$
2. $2 \text{ cytochrome } a_{3(ox)} + 2e^- \rightarrow 2 \text{ cytochrome } a_{3(red)}$
3. $1/2\, O_2 + 2H^+ + 2e^- \rightarrow H_2O$

$$\mathbf{cyt_c}\ \mathbf{E°} < \mathbf{cyt_{a3}}\ \mathbf{E°} < \mathbf{O_2}\ \mathbf{E°}$$

Conclusion: The reduction potential of the electron carriers, not their location in the mitochondrion, determines the flow of electrons. The higher the reduction potential of a compound, the more likely it is to accept electrons.

 ## What allows the mitochondrion to act as a galvanic cell?

A typical galvanic cell has two compartments linked by a metal wire that allows the transfer of electrons. In addition, in all electrochemical cells, the anode is the compartment where the oxidation occurs, and the cathode is where reduction occurs. The mitochondrion has a similar, though slightly modified, setup. In the mitochondrion, oxidation of NADH and FADH$_2$ occurs in the matrix. These electron carriers give off their electrons to complex I and complex II, respectively. Ubiquinone transfers these electrons to cytochrome b in complex III, which subsequently transfers them to cytochrome c. Cytochrome c exists partially in the intermembrane space. Therefore, the intermembrane space is where the reduction occurs, and is equivalent to a galvanic cell's cathode. Note that the sequestration of the electrons by the electron carriers, such as ubiquinone, holds a similar function as that of the wire in the galvanic cells. And, like galvanic cells, in each step of the ETC, the initial oxidation reaction is physically separated from the subsequent reduction reaction.

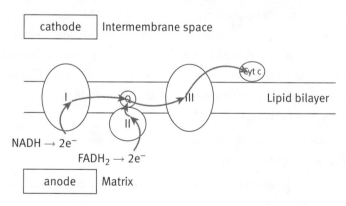

Takeaways

Double-check your work when you balance the cell equations to make sure that you haven't made any arithmetic errors. One small addition or subtraction mistake can have drastic consequences.

Things to Watch Out For

In a galvanic (voltaic) cell, the species with the higher reduction potential will be reduced. In an electrolytic cell, the species with the higher reduction potential will be oxidized.

 How can the EMF of the reaction below be calculated?

There are two equations for calculating the electromotive force of a reaction using its half-reactions' potentials. The equation on the left, which uses a negative sign, is used with the reduction potentials of both half reactions. Conversely, when using the equation on the right, we plug in the reduction potential of the compound that is reduced (at the cathode) and the oxidation potential of the compound that is oxidized (at the anode). The use of either equation, with the appropriate standard potentials, enable us to find the EMF.

$$2 \text{ cytochrome } c_{(ox)} + NADH \rightarrow 2 \text{ cytochrome } c_{(red)} + NAD^+ + H^+$$

$$2 \text{ cytochrome } c_{(ox)} + 2e^- \rightarrow 2 \text{ cytochrome } c_{(red)} \quad E° = +0.245 \text{ V}$$
$$NAD^+ + 2e^- + H^+ \rightarrow NADH \quad E° = -0.320 \text{ V}$$

$$E°_{cell} = E°_{cathode} - E°_{anode}$$

$$E°_{cell} = +0.245 \text{ V} - (-0.320 \text{ V})$$
$$E°_{cell} = +0.565 \text{ V}$$

$$E°_{cell} = E°_{reduction} - E°_{oxidation}$$

$$E°_{cell} = +0.245 \text{ V} + (+0.320 \text{ V})$$
$$E°_{cell} = +0.245 \text{ V} + (+0.320 \text{ V})$$
$$E°_{cell} = +0.565 \text{ V}$$

Related Questions

1. A galvanic cell is to be constructed using $MnO^{-4} \mid Mn^{2+}$ ($E°_{red} = +1.49$ V) and $Zn^{2+} \mid Zn$ ($E°_{red} = -0.76$ V) half-reactions in acidic solution. Assume that all potentials given are measured against the standard hydrogen electrode at 298K and that all reagents are present in 1 M concentrations. Calculate the $E°_{cell}$ for this cell and use that to determine how one could alter the cell setup to reverse the direction of current flow.

2. What would happen to the electromotive force produced by this cell if the amount of zinc present were doubled?

3. Given that the process of cellular metabolism via the ETC is used to "produce energy" in the form of energetic molecules and its reactions can be treated as those of a galvanic cell, what could be assumed about the electrochemical cell analogue of processes used to produce sugars or other storage molecules?

High-Yield Problem-Solving Guide questions continue on the next page. ▶ ▶ ▶

Solutions to Related Questions

1. Atomic Mass and Weight

1. A mole of any substance has a quantity of molecules equal to Avogadro's number ($N_A = 6.02 \times 10^{23}$ mol^{-1}). Therefore, while one mole of each substance has the same number of molecules, their masses will be different. Water has a molar mass of $2 \times 1 \frac{g}{mol} + 16 \frac{g}{mol} = 18 \frac{g}{mol}$, whereas carbon dioxide has a molar mass of $12 \frac{g}{mol} + 2 \times 16 \frac{g}{mol} = 44 \frac{g}{mol}$.

2. Given an atomic weight, it is not always possible to tell which isotope is most abundant; for example, bromine has two abundant isotopes: bromine-79 and bromine-81. The atomic weight sits between these two, at 79.9. Bromine-80, which is the closest to this value, is *not* the most abundant isotope. In this case, however, we can tell that silicon-29 must be the most common isotope. For the atomic weight to be 28.086, there must be enough silicon-29 to balance out the other isotopes, which are all less than 28.

3. This question is no different than the one given for zinc; here, let's use rounding to simplify the calculations. There are 200 total atoms, 40% of which are oxygen-16 and 30% of which are oxygen-14 and oxygen-18, respectively. If we multiply the atomic masses by their relative abundances and then add those results together, we get the atomic weight: $0.4 \times 16 \frac{amu}{atom} + 0.3 \times 14 \frac{amu}{atom} + 0.3 \times 18 \frac{amu}{atom} = 6.4 + 4.2 + 5.4 = 16 \frac{amu}{atom}$. The question asks for the molar mass, which is the mass of one mole of a substance in grams. This has the same value as the atomic weight, but different units. The molar mass of oxygen is $16 \frac{g}{mol}$. Note that we could have actually bypassed all of these calculations—the value reported in the Periodic Table *is* the atomic weight for an element.

2. Periodic Trends

1. In the neutral state, we would expect atomic radii to increase as we move from the upper right corner of the Periodic Table to the lower left: O < C < B < Be. When an atom gains electrons, its size increases significantly; in fact, we would predict that the divalent oxygen anion would be the largest. The opposite is true for cations; we would expect that boron, having lost three electrons, would be the smallest species. Therefore, we could rank these species from smallest to largest as follows:

    $$B^{3+} < C < Be < O^{2-}$$

2. Both phosphorus and selenium are equidistant from the upper right corner of the Periodic Table. However, the element with the higher effective nuclear charge (Z_{eff}) will have the higher ionization energy. Selenium has more electrons that shield valence electrons from the nucleus, reducing its Z_{eff}. Therefore, we would expect phosphorus to have a higher first ionization energy than selenium.

3. The second ionization energy of an element will always be higher than the first ionization energy, and the third ionization energy will always be higher still. Thus, this trend is expected. However, note that the increase from second to third ionization energy is much higher than the increase from first to second ionization energy. This implies that it requires some energy to remove one electron from the element, more energy to remove a second electron, and then much, much more energy to remove a third. In other words, this element seems to gain some stability after losing two electrons that is greatly disturbed by removing a third. Elements in Group IIA (Group 2) best fit this description; specifically, these are the ionization energies for strontium (Sr).

3. VSEPR Theory

1. The formula for aluminum trichloride is $AlCl_3$, and that for phosphorus tribromide is PBr_3. Based solely on the number of bonds, we might expect that both will have trigonal planar geometry. We must analyze the Lewis structure to determine if there are lone pairs. Aluminum is in Group IIIA (Group 13), so it has three valence electrons. Because it has already formed three single bonds, it has no electrons left to form a lone pair and will have the expected trigonal planar geometry. Phosphorus is in Group VA (Group 15), so it has five valence electrons. Because it has formed three single bonds, it must have one lone pair remaining; it therefore has trigonal pyramidal geometry. These two molecules vary in geometry because phosphorus has a lone pair, while aluminum does not.

2. A molecule with octahedral geometry has six groups organized around a central atom. In order to maximize the distance between any two groups, four of the groups occupy a square plane around the atom, and two are axial (up and down). This makes the maximum angle between two adjacent groups 90°. In a bent molecule, there are four groups (two of which are lone pairs). With fewer groups to arrange in three-dimensional space, each group can spread farther away from the others to minimize the amount of repulsion between adjacent groups. In bent molecules, this angle is around 104.5°. In other words, the angle is larger in a bent molecule than an octahedral molecule because fewer groups are sharing the same space.

3. The sulfate anion (SO_4^{2-}) has a tetrahedral structure because it has four groups arranged around a central atom (AX_4 geometry). The negative charges do not influence its overall geometry because the electrons are delocalized due to resonance. Thus, the charge is distributed around the whole molecule. Sulfuric acid (H_2SO_4) has two hydroxyl groups and two S=O bonds. While there are slight deviations in the angles between sulfate and sulfuric acid, the fact that both molecules arrange four groups around the central sulfur atom means that the geometry is tetrahedral for both.

4. Stoichiometry

1. Water is formed from hydrogen and oxygen gases through the reaction $2 H_2 + O_2 \rightarrow 2 H_2O$. 26 g of hydrogen gas corresponds to 13 moles of H_2 (molar mass = $2 \frac{g}{mol}$), which will produce 13 moles of water. 13 moles of water can then be converted to the various desired units to obtain 234 g H_2O (molar mass = $18 \frac{g}{mol}$) and 7.83×10^{24} molecules H_2O (using Avogadro's number). Note that the phrase *excess oxygen* implies that no limiting reagent calculation is required.

2. Let's start by balancing the reaction for the combustion of butane. The balanced reaction is:

$$2\ C_4H_{10} + 13\ O_2 \rightarrow 8\ CO_2 + 10\ H_2O$$

For this calculation, we are given the units of liters of gas at STP, which can be converted to moles using the equivalence factor $22.4\ \frac{L}{mol}$. Then, we'll use the mole ratio from the balanced equation above. Finally, we'll convert to a mass of butane using its molar mass ($4 \times 12\ \frac{g}{mol} + 10 \times 1\ \frac{g}{mol} = 58\ \frac{g}{mol}$). The calculations are:

$$179.2\ L\ CO_2 \times \left[\frac{1\ mol\ CO_2}{22.4\ L\ CO_2}\right]\left[\frac{2\ mol\ C_4H_{10}}{8\ mol\ CO_2}\right]\left[\frac{58\ g\ C_4H_{10}}{1\ mol\ C_4H_{10}}\right]$$

$$\approx \frac{180 \times 2 \times 60}{20 \times 8} = 9 \times 15 = 135\ g\ C_4H_{10}\ (actual = 116\ g)$$

3. This question is similar to those from before, but also introduces a limiting reagent. Start by balancing the equation. The balanced equation for the combustion of heptane is:

$$C_7H_{16} + 11\ O_2 \rightarrow 7\ CO_2 + 8\ H_2O$$

To determine the limiting reagent, convert each of the units given to moles: 44.8 L of heptane represents 2 moles of heptane; 1.204×10^{24} molecules of oxygen represents 2 moles of oxygen. Now, think logically. To react to completion, 2 moles of heptane would require 22 moles of oxygen based on the balanced chemical equation. Because we have only 2 moles of oxygen, this must be the limiting reagent. Now, we proceed as with any other stoichiometry calculation, using the 2 moles of oxygen as our given:

$$2\ mol\ O_2 \left[\frac{8\ mol\ H_2O}{11\ mol\ O_2}\right]\left[\frac{18\ g\ H_2O}{1\ mol\ H_2O}\right]$$

$$= \frac{288}{11} \approx \frac{286}{11} = 26\ g\ H_2O\ (actual = 26.2\ g)$$

5. Reaction Rates

1. The value of the rate constant can be determined by plugging the values from any of the trials into the rate law:

$$rate = k[HNO_2]^2$$
$$5.96 \times 10^{-6} = k[2.00 \times 10^{-2}]^2$$
$$5.96 \times 10^{-6} = k[4.00 \times 10^{-4}]$$
$$1.49 \times 10^{-2} = k$$

Its units can be determined by considering the other units in the rate law. The rate is measured in $\frac{M}{s}$, whereas $[HNO_2]$ is measured in M. Thus, the units of k must be $\frac{1}{M \cdot s}$.

2. If we compare trials 2 and 3, we can quickly determine that doubling the concentration of MeI while holding the concentration of pyridine constant resulted in a doubling of the rate. This means the exponent on MeI must be 1 ($2^1 = 2$). Determining the rate with respect to pyridine is a bit more challenging. Let's invent another trial based on what we know about MeI. If we hold the concentration of pyridine constant while *halving* the concentration of MeI, the rate should get cut in half:

Trial	$[C_5H_5N]$ (M)	$[MeI]$ (M)	Initial Rate $\left(\dfrac{M}{s}\right)$
1	1.00×10^{-4}	1.00×10^{-4}	7.50×10^{-7}
2	2.00×10^{-4}	2.00×10^{-4}	3.00×10^{-6}
2a	2.00×10^{-4}	1.00×10^{-4}	1.50×10^{-6}
3	2.00×10^{-4}	4.00×10^{-4}	6.00×10^{-6}

Now we can compare trials 1 and 2a. The concentration of pyridine doubles while the concentration of MeI remains constant. The rate also doubles. Therefore, the exponent on pyridine must also be 1 ($2^1 = 2$). The rate law is rate = $k[C_5H_5N][MeI]$. The rate constant could be calculated from trial 1:

$$\text{rate} = k[C_5H_5N][MeI]$$

$$7.5 \times 10^{-7}\,\frac{M}{s} = k[1.00 \times 10^{-4}\ M][1.00 \times 10^{-4}\ M]$$

$$75\ M^{-1}\,s^{-1} = k$$

3. The same steps can be used here, even though the numbers do not work quite "cleanly." Between trials 1 and 2, the concentration of cerium(IV) is constant while the concentration of iron(II) gets multiplied by just a bit more than 1.5. The rate also gets multiplied by just a bit more than 1.5. Therefore, the order with respect to iron(II) is 1 ($1.55^1 = 1.55$). Comparing trials 2 and 3, the concentration of cerium(IV) is multiplied by a bit more than 3, while the concentration of iron(II) is constant. The rate also gets multiplied by a bit more than 3. Therefore, the order with respect to cerium(IV) is also 1 ($3.1^1 = 3.1$). The rate law for the reaction of cerium(IV) and iron(II) is therefore rate = $k[Ce^{4+}][Fe^{2+}]$. The rate constant can be determined using trial 1:

$$\text{rate} = k[Ce^{4+}][Fe^{2+}]$$

$$2.00 \times 10^{-7}\,\frac{M}{s} = k[1.10 \times 10^{-5}\ M][1.80 \times 10^{-5}\ M]$$

$$2.00 \times 10^{-7}\,\frac{M}{s} \approx k[2 \times 10^{-10}\ M^2]$$

$$10^3\ M^{-1}\,s^{-1} \approx k\ (\text{actual} = 1.01 \times 10^3\ M^{-1}\,s^{-1})$$

6. Reaction Energy Profiles

1. The ratio of **B** to **A** at equilibrium can be determined using the equation $\Delta G° = -RT \ln K_{eq}$:

$$K_{eq} = \frac{[B]}{[A]} = e^{\frac{-\Delta G°}{RT}} = e^{\left(\frac{900\,\frac{cal}{mol}}{2\,\frac{cal}{mol \cdot K} \cdot 300\ K}\right)} \approx e^{1.5} \approx \text{between 3 and 9 (actual} = 4.56)$$

2. If a catalyst were added to the reaction of **A** forming **C**, the energies of **A** and **C** would not be changed. A catalyst only lowers the activation energy by lowering the energy of the transition state, and does not affect the energies of the reactants or products.

3. The complete reaction profiles for the reactions producing **B** and **C** from **A** are shown below. The intermediate leading to **C** is relatively more stable because the electrons in the double bond and in the lone pair on the oxygen atom can be delocalized. This can be seen in the plot from its lower energy.

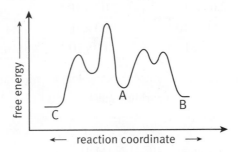

7. Hess's Law

1. An ideal hydrocarbon combustion reaction has two reactants (the hydrocarbon and oxygen) and two products (carbon dioxide and water). To determine the heat of reaction—which is combustion in this case—we will need the heats of formation for all of the reactants and products. We know that the heat of formation of diatomic oxygen gas is zero, so if we are given the heats of formation for carbon dioxide and water, we also need to know the heat of formation of ethane to determine the heat of combustion.

2. First, we must balance the chemical equation for the combustion of acetylene:

$$C_2H_2 + \frac{5}{2} O_2 \rightarrow 2\, CO_2 + H_2O$$

Note that the fractional stoichiometric coefficient in front of oxygen is not a concern because we'll be multiplying that coefficient by the heat of formation of oxygen gas, which is zero. We can apply Hess's law to answer this question. The heats of formation of all of the reactants and products are given, so we can solve for the heat of combustion:

$$\Delta H^\circ_{rxn} = \sum \Delta H^\circ_{f,products} - \sum \Delta H^\circ_{f,reactants}$$

$$\Delta H^\circ_{comb} = \left[2 \times \left(-393.5\, \frac{kJ}{mol}\right) + 1 \times \left(-241.8\, \frac{kJ}{mol}\right)\right] - \left[1 \times \left(226.6\, \frac{kJ}{mol}\right) + \frac{5}{2} \times \left(0\, \frac{kJ}{mol}\right)\right]$$

$$\approx 2 \times (-400) - 250 - 225$$

$$= -800 - 475 = -1275\, \frac{kJ}{mol}\ \left(actual = -1255.4\, \frac{kJ}{mol}\right)$$

3. Enthalpy is a state function; therefore, the path taken to get from the reactants to the products does not matter in determining the enthalpy change. If we add these five reactions together, the net reaction is:

$$Na\ (s) + \frac{1}{2}\, Br_2\ (g) \rightarrow NaBr\ (s)$$

This is equivalent to synthesizing sodium bromide from its elements in their standard states, which is, by definition, the standard heat of formation. Thus, the sum of the heats of reaction for these five reactions is $-359.9\, \frac{kJ}{mol}$.

8. Ideal Gases

1. Recall that nitrogen is a diatomic gas (molar mass $= 28 \frac{\text{g}}{\text{mol}}$) and that STP has conditions of 273 K and 1 atm (101.3 kPa). The given volume is 22.4 L and the ideal gas constant is $8.314 \frac{\text{J}}{\text{mol} \cdot \text{K}}$. Let's set up the ideal gas law equation again:

$$PV = nRT$$

$$n = \frac{PV}{RT} = \frac{(101.3 \text{ kPa})(22.4 \text{ L})}{(8.314 \frac{\text{J}}{\text{mol} \cdot \text{K}})(273 \text{ K})} = 1 \text{ mol}$$

One mole of diatomic nitrogen gas should have a mass of 28 g. A much simpler way of solving this question would require remembering the ideal gas conversion factor—1 mole of any ideal gas at STP occupies a volume of 22.4 liters, the volume given in the question.

2. We'll use the ideal gas law equation to solve this question, but first, make sure that all of the units match those of the ideal gas constant. We are given temperature in degrees Celsius. Remember to add 273 to this value to get the temperature in kelvins: 37°C + 273 = 310 K. Now we can plug into the ideal gas law equation:

$$PV = nRT$$

$$n = \frac{PV}{RT} = \frac{(1 \text{ atm})(0.5 \text{ L})}{(0.0821 \frac{\text{L} \cdot \text{atm}}{\text{mol} \cdot \text{K}})(310 \text{ K})} \approx \frac{0.5}{25} = 0.02 \text{ mol}$$

3. We can use the ideal gas constant equation again, but must convert the units of degrees Celsius to kelvins: 700°C + 273 = 973 K. Now we can plug into the ideal gas law equation:

$$PV = nRT$$

$$V = \frac{nRT}{P} = \frac{(2 \text{ mol})\left(62.4 \frac{\text{L} \cdot \text{mmHg}}{\text{mol} \cdot \text{K}}\right)(973 \text{ K})}{(4000 \text{ mmHg})} \approx \frac{2 \times 60 \times 1000}{4000} = 30 \text{ L (actual} = 30.4 \text{ L)}$$

9. Solution Equilibria

1. One can always use the same three-step method to determine molar solubility from K_{sp} or K_{sp} from molar solubility:

    1. Write out the balanced expression for the dissociation reaction
    2. Write the expression for K_{sp}
    3. Plug in the values from the question stem or data table

    If there is no common ion effect, this three-step method can be abbreviated by remembering three shortcuts. For a compound of general formula MX, $K_{sp} = x^2$, where x is the molar solubility; for a compound of general formula MX_2, $K_{sp} = 4x^3$; and for a compound of general formula MX_3, $K_{sp} = 27x^4$. When a common ion is present, these shortcuts cannot be used, and the three-step method listed above must be used instead.

2. Given a list of solubility constants, to find the substance most soluble in pure water, you would need to find the substance with the highest molar solubility. Because all of the compounds are being dissolved in pure water, the three shortcuts listed above can be used to speed up this process. Additionally, it makes most sense to compare the compounds with a particular general formula to one another, and then to compare the most soluble of the three groups. That is, of all the compounds with the general formula MX, the one with the highest K_{sp} is the most soluble. The same is true when comparing all of the compounds with the general formula MX_2 and those with the general formula MX_3. Finding the molar solubility of only these three salts and comparing them will allow you to find the most soluble salt as quickly as possible.

3. Decreasing the pH will cause greater dissolution of the sulfate salt because there will be more H^+ available to react with sulfate; this removes products from the dissociation reaction, driving the equilibrium toward the right. Increasing pH will have the opposite effect, driving the equilibrium to the left and disfavoring the dissolution of the sulfate salt.

10. Acids and Bases

1. The conjugate acid of a strong base can be assumed to have a K_a less than 10^{-14}. This extremely low K_a value indicates that the conjugate acid is inert, or nonreactive, as an acid.

2. The conjugate acid of a weak base can be assumed to have a K_a less than 1 but typically greater than 10^{-14}. This K_a value indicates that the conjugate acid is also weak, and will partially dissociate and react as an acid.

3. The K_a of an acid multiplied by the K_b of its conjugate base is 10^{-14}:

$$K_{a,\text{acid}} \times K_{b,\text{conjugate base}} = K_w$$

Taking the negative logarithm of both sides, we get:

$$pK_{a,\text{acid}} + pK_{b,\text{conjugate base}} = 14$$

Therefore, the pK_b of acetate is $14 - 4.76 = 9.24$.

11. Balancing Oxidation–Reduction Reactions

1. When present, transition metals are often the oxidized and reduced species in an oxidation–reduction reaction. After transition metals, elements further down in the *p*-block are relatively common as well. Therefore, let's start by identifying the oxidation states of zirconium and sulfur in this reaction. Zirconium starts with a +4 charge. Each of the hydroxide groups in $ZrO(OH)_2$ has a −1 charge, and oxygen usually has a −2 charge; so zirconium must have a +4 charge for the compound to be neutral overall. As a solid, zirconium has an oxidation state of 0. Therefore, zirconium has gained electrons and has been reduced—it is also the oxidizing agent. Sulfur, on the other hand, started with a +4 charge (each of the three oxygens has a −2 charge, and the overall charge of the molecule is −2) and ends with a +6 charge. Therefore, sulfur has lost electrons and has been oxidized—it is also the reducing agent.

2. We'll balance this oxidation–reduction reaction using the same method as the question. Note that no multiplication is needed in Step 4:

$$1. \begin{cases} PbSO_4 \rightarrow Pb + SO_4^{2-} \\ PbSO_4 \rightarrow PbO_2 + SO_4^{2-} \end{cases}$$

$$2. \begin{cases} PbSO_4 \rightarrow Pb + SO_4^{2-} \\ 2\,H_2O + PbSO_4 \rightarrow PbO_2 + SO_4^{2-} + 4\,H^+ \end{cases}$$

$$3. \begin{cases} PbSO_4 + 2\,e^- \rightarrow Pb + SO_4^{2-} \\ 2\,H_2O + PbSO_4 \rightarrow PbO_2 + SO_4^{2-} + 4\,H^+ + 2\,e^- \end{cases}$$

$$4. \begin{cases} PbSO_4 + 2\,e^- \rightarrow Pb + SO_4^{2-} \\ 2\,H_2O + PbSO_4 \rightarrow PbO_2 + SO_4^{2-} + 4\,H^+ + 2\,e^- \end{cases}$$

$$5. \quad 2\,PbSO_4 + 2\,H_2O \rightarrow Pb + PbO_2 + 2\,SO_4^{2-} + 4\,H^+$$

3. We'll balance this oxidation–reduction reaction using the same method as the question. Again, no multiplication is needed in Step 4:

$$1. \begin{cases} Zn \rightarrow Zn^{2+} \\ H_2O \rightarrow H_2 \end{cases}$$

$$2. \begin{cases} Zn \rightarrow Zn^{2+} \\ 2\,H_2O \rightarrow H_2 + 2\,OH^- \end{cases}$$

$$3. \begin{cases} Zn \rightarrow Zn^{2+} + 2\,e^- \\ 2\,H_2O + 2\,e^- \rightarrow H_2 + 2\,OH^- \end{cases}$$

$$4. \begin{cases} Zn \rightarrow Zn^{2+} + 2\,e^- \\ 2\,H_2O + 2\,e^- \rightarrow H_2 + 2\,OH^- \end{cases}$$

$$5. \quad Zn + 2\,H_2O \rightarrow Zn^{2+} + H_2 + 2\,OH^-$$

The amalgam must be kept dry because upon reacting with water in basic solution, hydrogen gas is generated. The expansion of the gas could potentially cause the crown or tooth to crack.

12. Electrochemical Cells

1. The EMF of this cell is +1.49 V – (–0.76 V) = 2.25 V. To reverse the direction of the current, one would have to place a battery of greater than 2.25 V in the circuit (with the positive terminal connected to the cathode of the galvanic cell setup). This would override the electromotive force of the cell, causing the current to flow the opposite direction.

2. The amount of zinc present would have no effect on the electromotive force of the cell (assuming that at least some zinc is present). The electromotive force is calculated as the difference in reduction potential between the cathode and the anode, and does not depend on the amount of zinc present.

3. Based on the information and process of logic used in the original question, it could be assumed that processes creating energy storage molecules function in some ways as a "reverse" of the process of events in carbohydrate metabolism. This would imply that, at at least some point in the production of sugars, EMF is negative and the reaction is driven forward by an outside energy source. This is analogous to an electrolytic cell, which requires energy input from a battery to drive the reaction forward.

5

Organic Chemistry

Key Concepts

Organic Chemistry Chapter 1

Nomenclature

Functional group priority

E Nomenclature

What is the IUPAC name for the following compound?

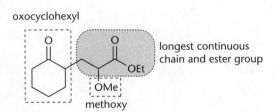

1 What is the longest carbon chain containing the highest-order functional group?

In this case, the highest-priority functional group is the ester. Therefore, we will name everything attached to the ester as a substituent, including the cyclohexyl ring on the left. The longest continuous chain containing the ester is three carbons long.

2 How would you number the longest carbon chain?

As mentioned in Step 1, the longest continuous chain containing the ester is three carbons long. Carbon 1 will be the carbonyl carbon because the ester is the highest-priority functional group. Because the ester has three carbons, this will be a *propanoate* ester.

3 How would you name the substituents?

The first substituent is the ethyl group on the ester, which we will identify by placing the word *ethyl* in front of the ester name. Next, there is a methyl group attached to an oxygen, which will be named as a *methoxy–* group. Finally, there is a cyclohexane containing a ketone, the naming of which will be discussed in the next step.

4 What numbers would be assigned to the substituents?

The ethyl group is the esterifying group and is not assigned a number, but rather is listed as an adjective before the rest of the molecule's name.

The methoxy group is on carbon 2.

How do we handle the ring attached to carbon 3? If there were nothing attached to the ring, we would simply name the ring as a *cyclohexyl* substituent. However, there is a ketone in the ring. Recall that when aldehydes or ketones are named as substituents, they are named with the prefix *oxo–*. The numbering works, as shown, by assigning the number 1 to the carbon attached to the parent carbon chain. Therefore, the ketone on the ring will be at carbon 2. We'll name the whole ring as a *(2-oxocyclohexyl)–* substituent and put it in parentheses to indicate that it follows a numbering system *within* a substituent.

5 What is the name of the compound?

The name of our compound will be ethyl 2-methoxy-3-(2-oxocyclohexyl) propanoate.

Takeaways

The key to the nomenclature problems is to be as systematic as possible. Don't try to do everything at once, or you risk confusing yourself.

Things to Watch Out For

Don't forget to include parentheses if a substituent is further substituted so that you don't confuse the two numbering systems.

Related Questions

1. How would the name be altered if the alkyl group attached to the ester oxygen contained substituents?

2. Upon reduction with sodium borohydride, followed by dilute acid workup, the molecule below gave two products in unequal yield. Draw them and provide the correct IUPAC name for each.

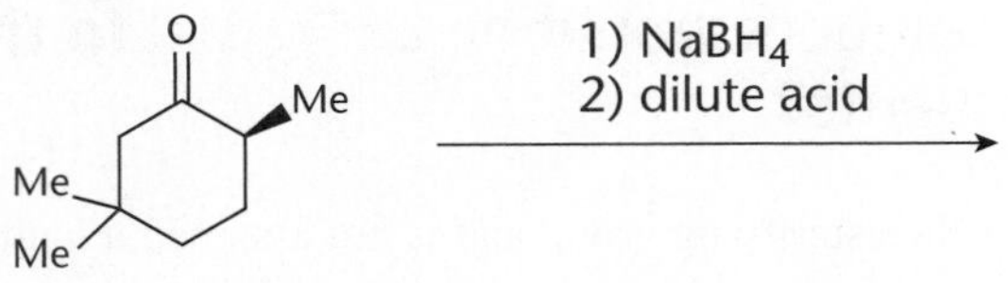

1) NaBH₄
2) dilute acid

3. What are the two possible products of the reaction shown below? Draw and provide IUPAC names for both.

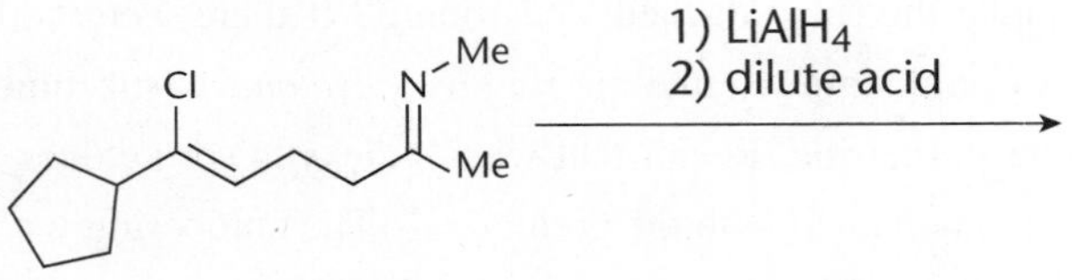

1) LiAlH₄
2) dilute acid

High-Yield Problem-Solving Guide questions continue on the next page. ▶ ▶ ▶

Key Concepts

Organic Chemistry Chapter 2

Stereochemistry

Fischer projections

Oxidation–reduction reactions

meso compounds

Ⓢ Stereoisomers

Ibuprofen (isobutylphenylpropionic acid), known widely by its trade name Advil, is a non-steroidal anti-inflammatory drug (NSAID) taken to alleviate pain. Only one of its two isomers is biologically active. In fact, the active isomer was once isolated and marketed but was quickly withdrawn due to its adverse side effects: consumers failed to account for its increased potency and took too much of it. Ibuprofen is currently produced industrially as a mixture of both isomers.

Isobutylphenylpropionic acid

Interestingly, *in vivo* testing revealed the existence of an isomerase, AMACR, able to convert 50–65% of the inactive isomer to the active one via a bond-breaking mechanism. Where does AMACR act on the ibuprofen molecule, and besides their biological activity, how do these two isomers differ?

❶ What are the defining characteristics of the three major classes of isomers?

There are three classes of isomers: conformational, configurational, and constitutional. Conformational and configurational isomers are both stereoisomers: these molecules have the same molecular formula and bond-to-bond connectivity but differ in the three-dimensional orientations of their atoms in space. Conformational isomers can be interconverted exclusively by rotations about their single bonds. For this reason, they are also referred to as rotamers. Conversely, configurational isomers cannot be converted into one another by rotation around a single bond. Lastly, constitutional, or structural, isomers are only related by the fact that they share the same molecular formula. They differ in their connectivity. Structural isomers often carry different functional groups.

Decreasing similarity

→

The <u>3C</u> isomers (**c**onformational, **c**onfigurational, **c**onstitutional)		
3D (stereo) isomers ("*conf*-isomers")		**Constitutional**
Conformational	**Configurational**	
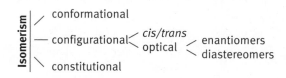		
Bond-breaking? Not required	**Bond-breaking?** Required	**Bond-breaking?** Required
Connectivity? Same	**Connectivity?** Same	**Connectivity?** Different

Takeaways

Any time optical activity is lost in a reaction, either a racemic mixture or an achiral compound has been formed.

Things to Watch Out For

With molecules that contain multiple stereocenters, consider drawing them as Fischer projections to be able to spot planes of symmetry or assign (*R*) and (*S*) nomenclature easily.

Isomerism
- conformational
- configurational ⟨ cis/trans ⟨ enantiomers / diastereomers, optical
- constitutional

② How can we eliminate the options of conformational and constitutional isomers in this case?

Conformational: The question tells us that an enzyme, AMACR, is necessary to convert the inactive isomer to the active one. Rotation around a bond would not require an enzyme catalyst that breaks bonds! This excludes the possibility of ibuprofen molecules being conformational isomers.

Constitutional: The molecule of ibuprofen is shown. Because both isomers can be represented using the same molecular diagram, both isomers must share the same functional groups and bond-to-bond connectivity. This observation eliminates the option of constitutional (structural) isomers.

3 Are the double bonds in ibuprofen a potential source of *cis/trans* isomerism?

Having determined that the ibuprofen isomers are configurational, we need to consider the possibility of either optical or *cis/trans* (geometric) isomerism. *Cis/trans* isomerism usually relies on orientation of functional groups about double bonds. There are two types of double bonds in the ibuprofen molecule: those in the benzene ring and the one in the carbonyl group. Let's consider the benzene ring first: because there are no adjacent substituents on the ring, the ring itself does not constitute a source of *cis/trans* isomerism.

Regarding the carbonyl carbon, it is clear that the oxygen does not have any substituents. In order to have *cis/trans* isomerism, two sp^2 carbons must be bonded via a double bond and each carry two different substituents. Given that this criterion is not satisfied in the ibuprofen molecule, its double bonds are not a source of *cis/trans* isomerism. So, by process of elimination, the ibuprofen isomers must exhibit some form of optical isomerism.

trans-2-butene   *cis*-2-butene

4 Why are the two isomers of ibuprofen enantiomers and not diastereomers?

The hallmark of both enantiomers and diastereomers is that they possess at least one chiral center. A chiral carbon is an sp^3 hybrid carbon bound to four different substituents. Recall that diastereomers are compounds with at least two stereocenters that have different configurations at one or more (but not all) stereocenters. Because ibuprofen only has one chiral center (the α-carbon of the carbonyl group), ibuprofin's isomers must be an enantiomers. As this α-carbon is the location that dictates the identity of the isomer, it must be the site of action of AMACR.

5 How does the active isomer differ from the inactive isomer?

Enantiomers only differ in two ways: they rotate plane-polarized light in opposite directions and they react differently in chiral environments. For example, "dex" ibuprofen, the active isomer, rotates light clockwise; it's enantiomer, the inactive ibuprofen isomer, rotates light counterclockwise. But also, within your body, enzymes that are reactive to dexibuprofen are not reactive to its enantiomer. This difference stems from the fact that enzymes are a form of "chiral environment." Put another way, when it comes to reactions with enzymes, the enantiomer that you use matters!

Related Questions

1. The specific rotation of a certain D-carbohydrate is +140°. What is the specific rotation of its L-isomer? What would be the observed rotation in a solution containing 1 M of each isomer?

2. Which of these three D-aldopentoses (shown below) would result in achiral polyols when subjected to borohydride reduction?

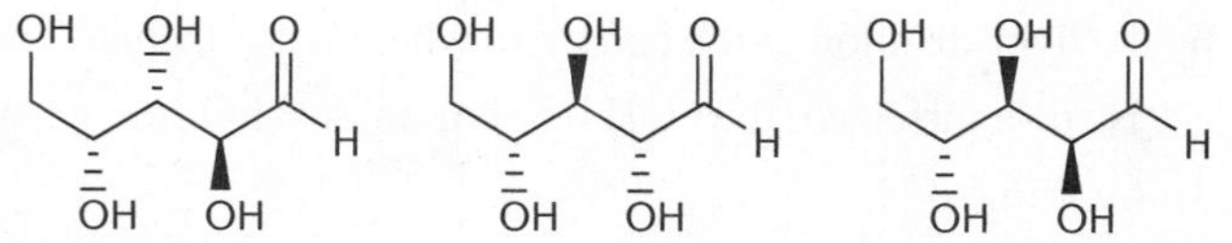

3. A *meso* compound contains chiral centers but is overall achiral. Is the reverse true? That is, can a molecule that does *not* contain chiral centers be overall *chiral*?

⊟ Hybridized Orbitals

Orbital hybridization affects more than just the shape of molecules. In fact, *sp* hybrid orbitals can have quite different properties than sp^3 orbitals, even on atoms of the same element. The compounds below differ only in the hybridization of the carbon atom highlighted. First, arrange them in order of increasing pK_a for the highlighted protons. Then, arrange them by increasing bond dissociation energy between the two carbons.

① How do the pK_a values of the compounds compare?

The K_a (acid dissociation constant) of a molecule measures its acidity; the higher the K_a, the more acidic the compound. The term "p" is shorthand for −log, so pK_a = −log K_a. This mathematical relationship means that the *lower* the pK_a, the more acidic the compound. In each of these three compounds, the proton to be lost differs only by the hybridization of the carbon to which it is attached. The C–H bond in compound **1** is sp^3-hybridized, the C–H bond in **2** is sp^2-hybridized, and the C–H bond in **3** is *sp*-hybridized.

To understand the acidity trend in this group, we need to consider not only the neutral molecules but their anions as well. Each anion in this case has a lone pair of electrons where the proton was once bound. In the case of compound **1**, the electrons are in an sp^3 orbital; in **2**, in an sp^2 orbital; and in **3**, in an *sp* orbital.

It's useful to say that these lone pairs are in different hybrid orbitals, but how does that actually affect the relative stabilities of their anions? The anions' stabilities are dependent on the energies of the two electrons in the lone pair; these energies are in turn based on the energies of the orbitals in which they are located. Each of these orbitals is a hybrid of *s*- and *p*-orbitals. Remember that *s*-orbitals are much lower in energy than their *p* counterparts, and hybrid orbitals have energies that are weighted averages of their constituent orbitals. Therefore, the anion with the most *s* character should be the most stable, and the anion with the most *p* character should be the least stable. The order of anion stability is:

1 < 2 < 3

This isn't the whole picture, though. We were asked to arrange the compounds in order of increasing pK_a, not anion stability—and the less stable the anion is, the less acidic its conjugate acid will be. The less acidic the conjugate acid, the higher the pK_a. Putting it all together, the molecules can be arranged in order of increasing pK_a:

3 < 2 < 1

2 **How do the bond dissociation energies of the carbons in the 3 compounds compare?**

Bond dissociation energy is the energy required to break a bond, leaving one electron with each fragment, which forms two radicals. It is the standard measure of bond strength in organic chemistry. In general, bond strength and bond energy are related: as bond strength increases, so does bond dissociation energy. Also, the shorter the bond, the stronger it is. For instance, triple bonds are stronger and shorter than double bonds between the same atoms; these, in turn, are shorter and stronger than single bonds. Thus, the C≡C bond in the *sp*-hybridized compound (**3**) should be the shortest and strongest. The weakest will be the one between two *sp*^{3}-hybridized carbons (**1**). Thus, the order should be:

1 < 2 < 3

Takeaways

Stronger bonds are shorter bonds: triple bonds are the strongest and shortest, while single bonds are the longest and weakest.

Things to Watch Out For

pK_a (acid strength) is determined not only by the acid itself but also by the conjugate base that is formed upon the loss of a proton.

Related Questions

1. Which will lose a hydrogen more easily, a compound with a higher or lower pK_b?

2. What is the hybridization of the carbon atom in carbon dioxide?

3. What is the highest bond order in 5-hydroxy-2,6-dimethyl-3-heptanone?

E Nucleophilicity Trends

Rank the following compounds in order of increasing nucleophilicity toward the same electrophile in a polar protic solvent:

$$CH_3OH \qquad Et_3N \qquad H_3C-C(=O)-O^{\ominus} \qquad Et_3P \qquad CH_3O^{\ominus}$$

1 Which nucleophiles have the same attacking atom, and how do they compare?

Let's look at the oxygen-containing nucleophiles first. Here, the methoxide anion is more basic than the acetate anion $\left(CH_3CO_2^-\right)$, which in turn is more basic than methanol. With the methoxide anion, the lone pair on oxygen is "stuck" on the oxygen atom, whereas with acetate, the negative charge can be delocalized through resonance; this makes methoxide more basic. Both molecules are more basic than methanol because methanol lacks a negative charge. When comparing nucleophiles with the same attacking atom, nucleophilicity parallels basicity. Therefore, the methoxide anion will be the most nucleophilic of the three oxygen-containing molecules.

2 Which nucleophiles have attacking atoms in the same group?

Because phosphorus is directly below nitrogen in the Periodic Table, triethylphosphine is more nucleophilic than triethylamine. This is where the nature of the solvent makes a big difference. The more basic molecules are better hydrogen bond acceptors, which means that they will be surrounded by the polar protic solvent molecules and therefore less available to attack the electrophile. The differences in basicity are less pronounced when comparing attacking atoms that are the same or are in the same *period*; this effect is only notable when the attacking atoms are in the same *group*.

Comparing the basicity of triethylphosphine and triethylamine is a bit more complicated than the previous molecules. The key to determining their relative basicities is remembering that in triethylphosphine, the lone pair on phosphorus is contained in an sp^3-hybridized orbital that is made up of one $3s$ and three $3p$ orbitals. Contrast this with triethylamine, where the nitrogen lone pair is in an sp^3-hybridized orbital composed of one 2s and three 2p orbitals. The hybridized orbitals on phosphorus are composed of larger atomic orbitals, and as such are larger than the hybridized orbitals on nitrogen.

This, in turn, means that the electrons in the phosphorus lone pair are more stable because they have a larger volume of space in which to move around. If the phosphorus lone pair is more stable, then it is less basic (less likely to reach out and grab a proton). Therefore, triethylphosphine is a better nucleophile in a polar protic solvent than triethylamine. Again, this phenomenon is only relevant when the attacking atoms of two nucleophiles are in the same group.

If the solvent were polar aprotic, then the trend would be exactly the opposite. Here, the hydrogen bonding effect is removed, so the molecules with the most localized charge density—the most basic—will also be the most nucleophilic.

3 What relationship exists between nucleophiles in the same period?

Now the question is between the two groups we have ordered separately. Which one is more nucleophilic? In most cases, this question is answered by realizing that for different nucleophiles where the attacking atoms are in the same period, nucleophilicity roughly parallels basicity. Nitrogen is less electronegative than oxygen. The less electronegative an atom, the less pull it will have on its attached electrons, making those electrons more available to act as a basic pair. Therefore, nitrogen is more basic than oxygen, and thus triethylamine is more nucleophilic than the oxygen-containing compounds. In summary, the compounds can be ranked by nucleophilicity toward the same electrophile in a polar protic solvent:

$$CH_3OH < CH_3CO_2^- < CH_3O^- < Et_3N < Et_3P$$

This trend is borne out experimentally. The relative reactivities of each nucleophile toward CH_3I in CH_3OH are as follows:

Nucleophile	Relative Rate
CH_3OH	1
$CH_3CO_2^-$	20,000
CH_3O^-	1,900,000
Et_3N	4,600,000
Et_3P	520,000,000

Takeaways

Sort out differences in nucleophilicity two or three molecules at a time. Don't try to order all the molecules at once or you risk confusing yourself.

Things to Watch Out For

Remember that basicity sometimes *parallels* nucleophilicity (comparing nucleophiles with the same attacking ion, in the same period, or in polar aprotic solvents) but sometimes opposes it (comparing nucleophiles in the same group in a polar protic solvent).

Related Questions

1. Place the following molecules in order of increasing nucleophilicity: pyridine (benzene with one of the carbons in the ring replaced by a nitrogen), triethylamine, acetonitrile (CH_3CN), and 4-dimethylaminopyridine (DMAP). (Note that the solvent doesn't impact nucleophilicity here because the same atom is nucleophilic in all four compounds.) Which of the two nitrogens in DMAP is more nucleophilic, and why?

2. How would the nucleophilicity of fluoride, chloride, bromide, and iodide rank in an S_N2 reaction with methyl iodide in methanol? In dimethyl sulfoxide (DMSO)?

3. How would you order the nucleophilicity of the following molecules in methanol: Et_3N, Ph_3P, Et_3P, Ph_3N, and Et_3As? Provide a rationale for your ordering.

High-Yield Problem-Solving Guide questions continue on the next page. ▶ ▶ ▶

E Reactions of Alcohols

Determine the identities of products 3 and 4. What is the relationship between the two products?

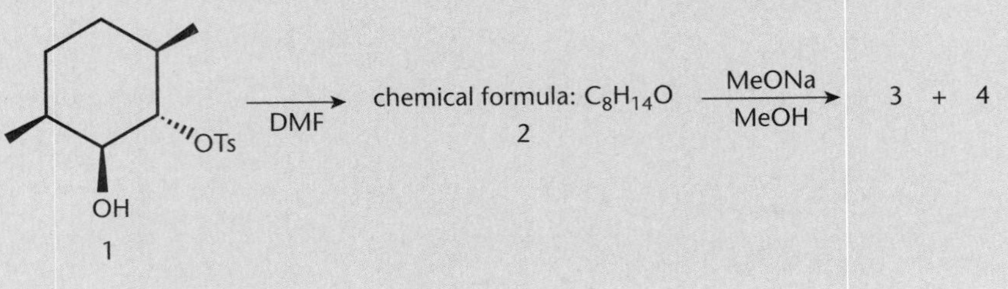

$$\xrightarrow{\text{DMF}} \quad \text{chemical formula: } C_8H_{14}O \quad \xrightarrow[\text{MeOH}]{\text{MeONa}} \quad 3 \ + \ 4$$

2

① What would be the structure of the intermediate, 2?

Before **3** and **4** can be identified, the structure of **2** must be found. Interestingly, no reagents are given in this reaction aside from the solvent (DMF), so we will have to look more closely at the starting reactant. The molecule we are given has two reactive groups that immediately stand out: –OTs, a tosylate group, and –OH, a hydroxyl group. Tosylate is a very good leaving group; the hydroxyl substituent is a poor leaving group and a good (but not great) nucleophile.

Because we have a leaving group but no strong base or nucleophile, the reaction is likely to proceed through an S_N1 mechanism (or E1, although elimination reactions are not tested directly on the MCAT). The first step is the loss of the leaving group, leaving a carbocation intermediate, as shown below:

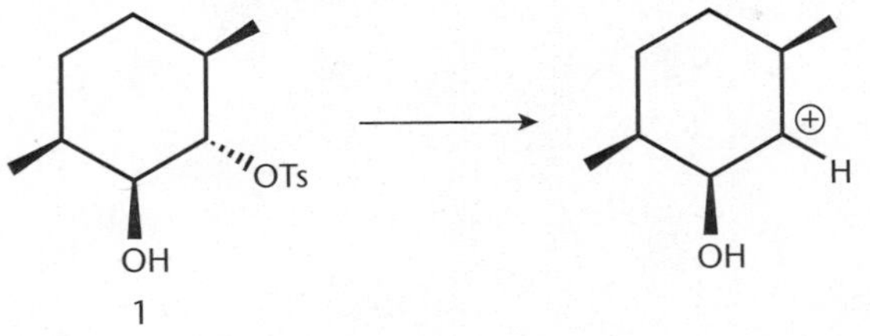

This carbocation intermediate has the formula $C_8H_{15}O$—that's just one hydrogen more than what we're given for intermediate **2**. What's the next step? Next, the nucleophilic hydroxyl group can attack the carbocation. This still leaves us with an intermediate containing an extra hydrogen, which the tosylate group can remove, giving us the structure of intermediate **2**.

2   What would be the structures of 3 and 4?

Starting with the structure that we figured out for **2**, we can work through the next set of reagents and determine products **3** and **4**. Our intermediate, **2**, only has one particularly reactive group, the epoxide. While epoxides are not on the official MCAT content lists, it makes sense that the oxygens they contain make good leaving groups. In **2**, the three-membered oxygen-containing ring has a high degree of angle strain, which can be relieved by opening the ring.

Epoxides can open under acidic or basic conditions, but don't do much else. Looking at the reactants, we have sodium methoxide in methanol. Methoxide is both a good base and a good nucleophile. In this case, the less reactive methanol will act as the solvent.

Methoxide can attack at either of two positions, performing our predicted ring-opening reaction:

Protonating these molecules gives us our final products, **3** and **4**. The overall reaction is shown below:

Takeaways

Reactivity in the presence of a good leaving group depends on the quality of the nucleophile. Identifying the quality of a leaving group and the strength of the nucleophile will help determine what happens during the reaction.

Things to Watch Out For

Be sure to keep in mind the stereochemistry of a given reaction; backside attack and planar intermediates both indicate specific changes to stereochemistry. Know how to distinguish between different types of isomers.

3 What is the relationship between 3 and 4?

Because **3** and **4** share a molecular formula, they are isomers of one another. These molecules have the same connectivity, so they are stereoisomers—not configurational isomers:

While the connectivity is the same, the stereochemistry is different. These two molecules are nonsuperimposable mirror images, and are therefore enantiomers.

Related Questions

1. How would the reaction be affected by changing the hydroxyl group to a stronger nucleophile?

2. How would the reaction be affected by changing the tosylate to a hydroxyl group?

3. Is it possible to favor one product over the other in this reaction?

High-Yield Problem-Solving Guide questions continue on the next page. ▶ ▶ ▶

Key Concepts

Organic Chemistry Chapter 6

Reaction mechanisms

Oxidation–reduction reactions

Aldehydes and ketones

Oxygen-containing functional groups

Carbonyl chemistry

Identifying the Structure of an Unknown
E Oxy Compound

A student carried out the following series of transformations in the lab:

OH

PCC

(S)-3-methylcyclohexanol

A
$C_7H_{12}O$

cat. H_2SO_4
$PhCH_2NH_2$

B + **C**
$C_{14}H_{19}N$

1) $LiAlH_4$
2) dilute acid

D + **E**
$C_{14}H_{21}N$

Upon **A**'s reaction with catalytic acid and benzylamine, the student obtained a mixture of two products, **B** and **C**. The mixture of **B** and **C** was subjected to lithium aluminum hydride. Following a dilute acid workup and chromatographic separation, two more products, **D** and **E**, were obtained in unequal yield.

Identify the structures of compounds **A** through **E**, given their molecular formulas.

① **What is the product of reaction of this alcohol with PCC?**

The first step of this process is to take the secondary alcohol and subject it to pyridinium chlorochromate (PCC) oxidation, which results in a ketone:

OH

PCC

O

Me

Me

$C_7H_{12}O$

2   What is the identity of product B?

You should suspect that some sort of nucleophilic addition is going to take place. Recall that a ketone that reacts with a primary amine gives an imine:

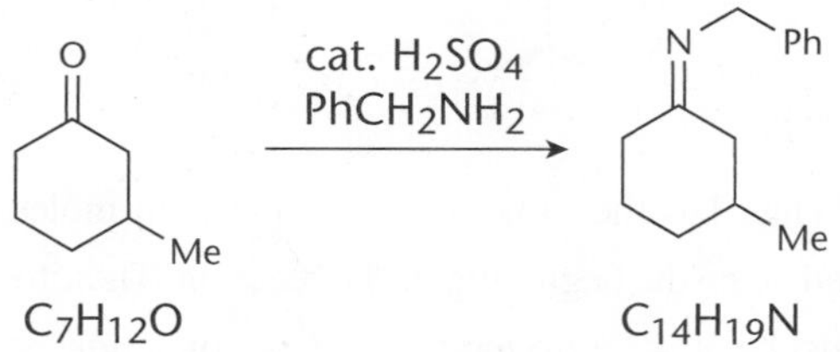

$$C_7H_{12}O \qquad\qquad C_{14}H_{19}N$$

3   What must be the identity of the other product, C?

Remember that any time there is an immovable bond (like a double bond), there exists the possibility of having *cis/trans* isomers. The immovable bond doesn't have to be a carbon–carbon double bond; it can also be between different constituent atoms, like the imine here. Thus, the other product must be the other geometric isomer:

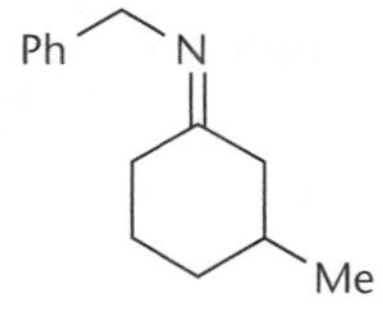

4   What are the structures of products D and E?

Reacting each of the imines with lithium aluminum hydride (LiAlH$_4$ or LAH) will result in an amine. Note that you should be able to identify LiAlH$_4$ as a reducing agent because it contains a large number of hydrides (H$^-$) attached to a metal (aluminum):

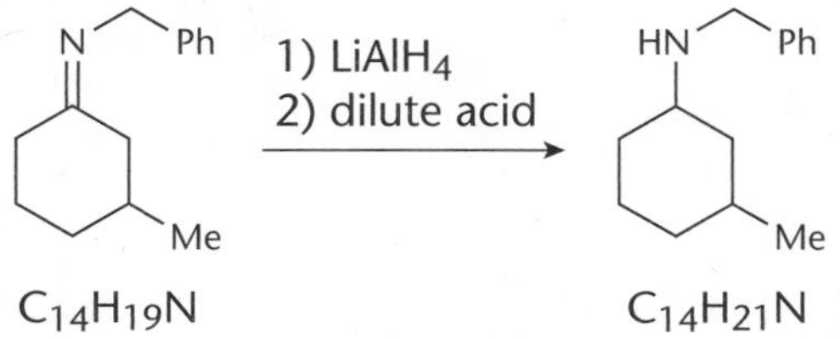

$$C_{14}H_{19}N \qquad\qquad C_{14}H_{21}N$$

Once reduced, the carbon atom from the imine can assume one of two different stereochemistries, resulting in two possible products:

The methyl stereocenter has the same orientation in both molecules because it has remained unchanged since the beginning of the reaction. Therefore, the difference in stereochemistry must be at the nitrogen-bearing carbon atom.

Related Questions

1. How many stereoisomers exist for product **E** (whether or not they were obtained in this particular reaction)?

2. The final amine products were mixed, in equal amounts, with an achiral solvent. The solution was placed in a polarimeter and found to rotate plane-polarized light. Why is this finding not surprising?

3. In addition to compounds that contain double bonds, what other class of compounds can have *cis/trans* isomers?

High-Yield Problem-Solving Guide questions continue on the next page. ▶ ▶ ▶

Key Concepts

Organic Chemistry Chapter 7

Enolate chemistry
Aldehydes and ketones
α-Hydrogen acidity
Keto–enol tautomerization

⊟ Enolate Chemistry

Enolate chemistry has blossomed in the last few decades, becoming one of the most powerful tools in organic synthesis. In reactions involving enolates, small structural differences in reactants can result in vastly different products. Take, for instance, the reaction of cyclohexanone, a common synthetic precursor, with methyl vinyl ketone; what are the products of this reaction?

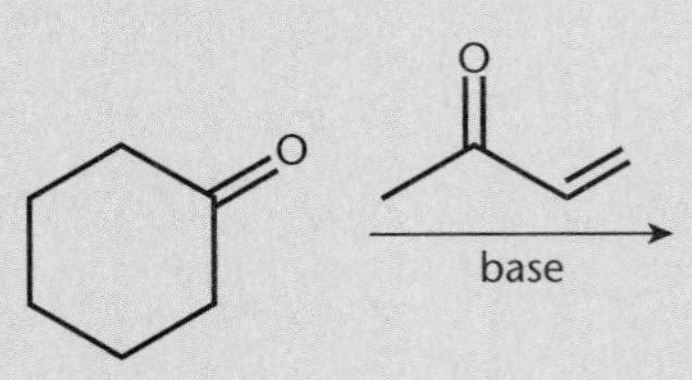

![Reaction scheme: cyclohexanone plus methyl vinyl ketone with base]

---

1 ## What is the enolate form of the reactant?

Whether you recognize the reaction of cyclohexanone and methyl vinyl ketone as what is called a Robinson annulation, you should recognize that the use of ketones under basic conditions means we'll be doing enolate chemistry. Whenever you see a base and an aldehyde or ketone, look for the most acidic proton and deprotonate. In this case, the most acidic protons are on the cyclohexanone. Deprotonating one of the α-carbons results in the cyclohexyl enolate:

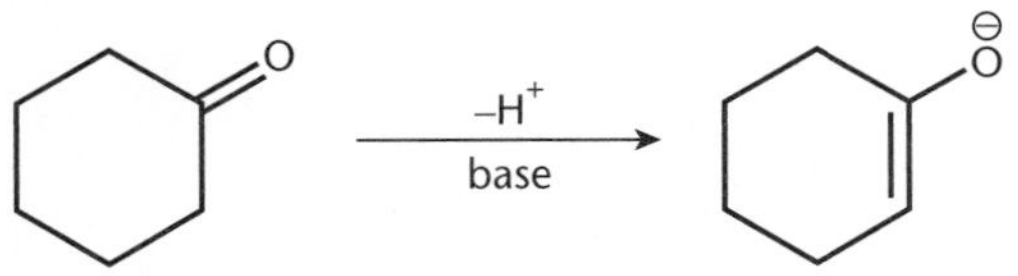

![Cyclohexanone converting to cyclohexyl enolate with –H+ / base]

---

2 ## What is the product of the enolate reacting with methyl vinyl ketone?

Enolates are nucleophilic species. Therefore, the next step will require an electrophile. The only electrophile present in this reaction is the methyl vinyl ketone. It would be reasonable to assume that the enolate will attack the methyl vinyl ketone at the carbonyl carbon, its most electrophilic region, as shown below:

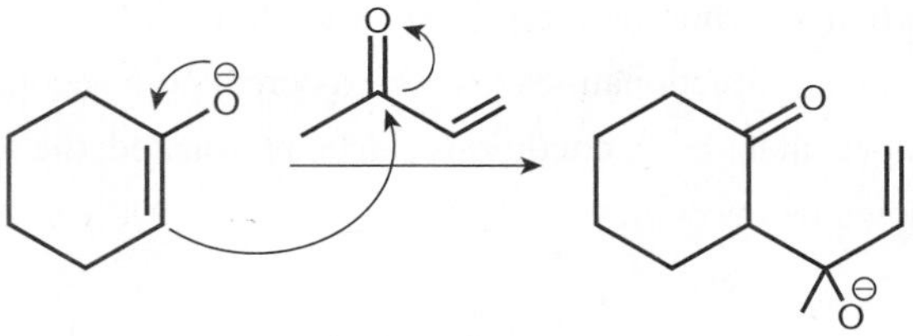

However, this requires a very crowded transition state to form. Instead, the enolate could attack C-4 on the methyl vinyl ketone because the primary carbon is significantly easier to reach. Indeed, this is what happens, resulting in the enolate shown below:

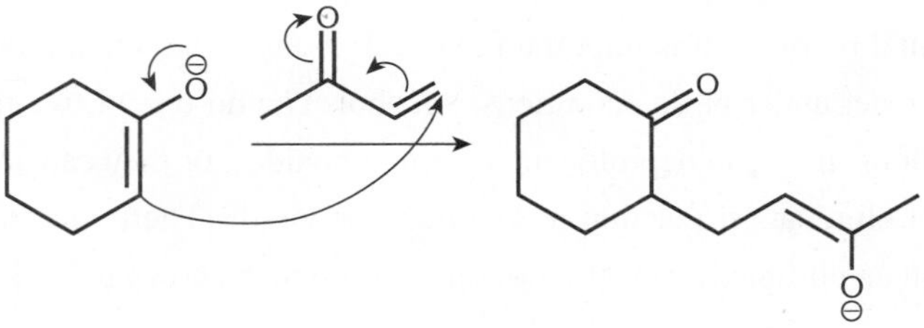

③ How would the product enolate react with an electrophile?

Again, we're left with an enolate and a carbonyl. In theory, we could have the enolate attack the carbonyl immediately. However, this results in the rather unfavorable formation of a four-membered ring. The cyclization is significantly slower than the isomerization of the enolate, shown below.

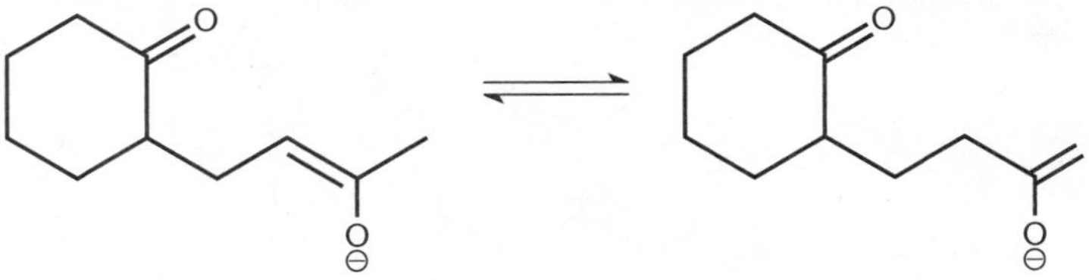

This isomerization is important because the newly formed primary enolate can attack the carbonyl carbon and form a six-membered ring—much more favored than the formation of a four-membered ring. It should be noted that the isomerization is a multistep acid–base reaction, involving protonation of one α-carbon, followed by deprotonation of the other.

Attack of the carbonyl carbon by the primary enolate is shown below. This results in alkoxide and ketone functionalities, as well as creating a second ring structure. Generally, alkoxides under basic conditions can be protonated; the result in this case is the alcohol shown below:

Is this our final product? It is important to check if any of the remaining functional groups will react under basic conditions. Alcohols can do one of two things under basic conditions: they can deprotonate to form alkoxides, or they can act as leaving groups in an elimination reaction to form double bonds. There are three possible isomers from an elimination reaction of our last product, shown below:

Takeaways

Enolates can allow the formation of cyclic products in molecules that contain both a carbonyl and the enolate functionality.

Things to Watch Out For

Keep track of nucleophiles and electrophiles to understand how the reaction will progress. Make sure you've reached the end of the mechanism and that nothing else will react!

At first, one might assume that the tetrasubstituted double bond (the middle structure) would be preferentially formed because more-substituted double bonds tend to be more stable than less-substituted ones. However, the rightmost compound, while having only three substituents on the double bond, also puts the double bond in conjugation with the carbonyl. This conjugation trumps the extra substitution, so elimination of the alcohol will result preferentially in the final product, shown below:

Related Questions

1. Cyclohexanone does not have a proton present on its oxygen atom. How does the first step of this reaction mechanism proceed?

2. What does it mean for a molecule to be conjugated? Why does conjugation increase the stability of a molecule?

3. When 2,2,6,6-tetramethylcyclohexanone reacts with 4,4-dimethylhexanal, which molecule serves as the nucleophile? Which molecule serves as the electrophile?

S Nucleophilic Acyl Substitution

Carboxylic acid derivatives are found in a vast array of organic molecules. Luckily, these can often be prepared from a few common intermediates. Take, for instance, the following reactions with methyl pivalate. Each of these produces a unique product. What are the products of each of these reactions?

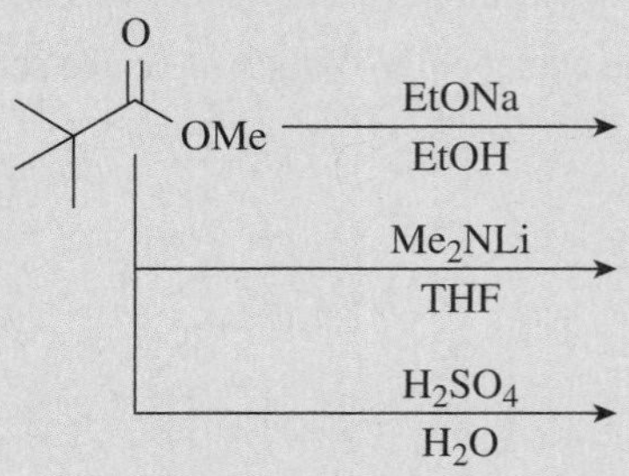

Key Concepts

Organic Chemistry Chapter 8
Nucleophilic acyl substitution reactions
Carbonyl chemistry
Carboxylic acids
Carboxylic acid derivatives

1 **What functionality does each of the carboxylic acid derivatives have in common?**

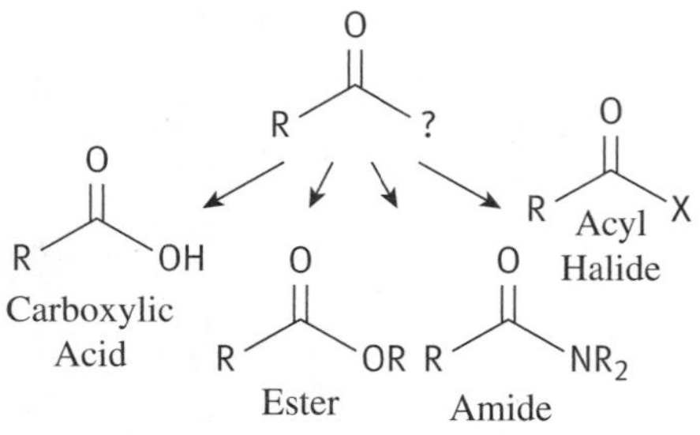

Carboxylic acid derivatives all contain a carbonyl carbon double bound to the more electronegative element oxygen. The uneven sharing of electrons in this bond ensures that in all carboxylic acid derivatives, the carbonyl carbon is electrophilic.

2 **Using the general mechanism for nucleophilic acyl substitution, sketch the transesterification reaction.**

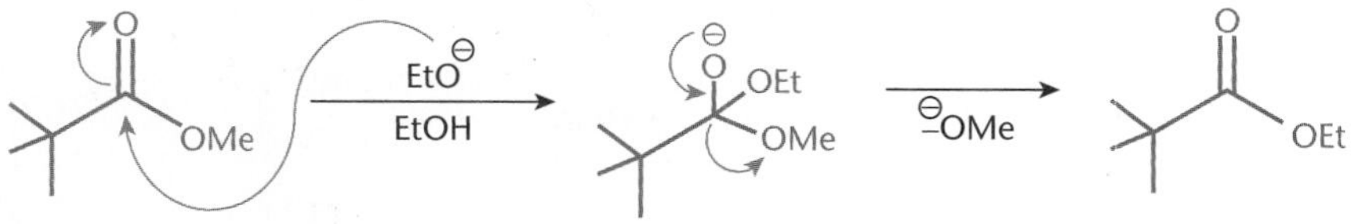

In the first reaction, the ethoxide anion is a strong nucleophile capable of attacking the electrophilic carbonyl carbon. The first step in this reaction is an attack on the carbonyl carbon by ethoxide. The intermediate formed after the nucleophilic attack is a tetrahedral alkoxide. The lone pair of electrons on the oxygen can push back down and reform the carbonyl if there is a good leaving group. Both ethoxide and methoxide are reasonable leaving groups under basic conditions. Displacing ethoxide would simply return the original compound; to get a new product, the methoxide must leave to give an ethyl ester as the final product.

3 How is the amide formation reaction similar to the transesterification reaction?

When you treat an ester with a nitrogen-containing anion as opposed to an alkoxide, the overall reaction mechanism is similar. However, this process now forms an amide. The nitrogen-containing anion is extremely nucleophilic, much like the alkoxide anion. In fact, the electron pushing is identical. In this case, the reaction is driven by the inherent stability of the final product. Amides, such as the final product shown, are relatively robust and nonreactive compounds.

4 Why does hydrolysis require an acid catalyst?

In both of the previous reactions, strong nucleophiles were able to drive the system toward product formation. Water is substantially weaker as a nucleophile, so in order to facilitate attack, the carbonyl carbon must become more electrophilic. The first

Takeaways

No matter the reaction, identifying nucleophiles, electrophiles, and leaving groups will help you figure out how the reaction will progress.

Things to Watch Out For

Remember if you see an acidic aqueous workup to keep an eye out for the possibility of hydrolysis—particularly in esters.

step in acidic conditions is protonation of the most basic part of the molecule. In this case, the most basic part is the carbonyl oxygen. Once the oxygen is protonated, the carbonyl carbon becomes significantly more electrophilic, allowing attack from the relatively non-nucleophilic water.

Related Questions

1. Which is a better leaving group, methanol or ethanol?

2. What is a lactone? What reactant(s) would be required to make a lactone through a nucleophilic acyl substitution mechanism?

3. What is a lactam? What reactant(s) would be required to make a lactam through a nucleophilic acyl substitution mechanism?

High-Yield Problem-Solving Guide questions continue on the next page. ▶ ▶ ▶

Key Concepts

Organic Chemistry Chapter 9

Anhydrides

Esters

Amides

Leaving group ability

Steric hindrance

⊟ Properties of Carboxylic Acid Derivatives

In an ideal reaction, there is exactly one nucleophile present that can react with exactly one electrophile. In practice, though, there are often multiple nucleophiles (and electrophiles) that compete with one another. Rank the carbonyl carbons in the following compounds in order of increasing electrophilicity toward a single nucleophile in a single polar protic solvent.

1 What is the order of the acyl derivatives by reactivity?

In the compounds listed above, there are three types of carbonyl functional groups: amides, esters, and anhydrides. These functional groups can be ranked by electrophilicity, which is based on the stability of their leaving groups. Anhydrides (**1** and **4**) are the most reactive (that is, most electrophilic) because their leaving groups are carboxylic acids, which can delocalize the lone pair left behind after heterolysis through resonance. Esters (**2** and **3**) are the next most reactive, while amides (**5**) are the least reactive. This is because the leaving group of an ester is an alkoxide ion or alcohol, which is less nucleophilic than the nitrogen-containing anion or amine leaving group of an amide. Leaving groups should be unreactive, so stronger nucleophiles generally make poorer leaving groups.

2 How would steric hindrance in the given molecules affect this reactivity ranking?

Next, we can rank within these groups. The more steric hindrance an electrophile presents to a nucleophile, the less reactive that molecule will be. Of the two anhydrides, **1** has significantly more steric hindrance, caused by its bulky phenyl groups, and is therefore less reactive than **4**, which has hydrogen atoms instead. Of the esters, the additional phenyl group on **3** means it will not react as readily as the less sterically hindered **2**.

3 What is the final order of reactivity?

With these rules in mind, we can order our compounds. The sterically hindered, least reactive amide **5** will be the least electrophilic; the unhindered and most reactive anhydride **4** will be the most electrophilic. **1**, **2**, and **3** fall in the middle. It is important to keep in mind that reactivity based on functional group may be counterbalanced by steric hindrance. In this case, although it is an anhydride, **1** will be significantly impeded by sterics, and will therefore be less reactive than the esters. On the MCAT, you would not be expected to know the relative contribution of each of these trends, but could be tested on any of them in isolation. This gives us our final order:

$$5 < 1 < 3 < 2 < 4$$

Related Questions

1. Why is it important that the peptide bonds in proteins, which are a special form of amide bonds, are relatively stable?

2. Which would you expect to be more reactive toward a given nucleophile, an aldehyde or a comparable ketone? Why?

3. Which carboxylic acid derivatives can participate in hydrogen bonding with themselves?

Takeaways

Ordering carboxylic acid derivatives by reactivity depends not only on the identity of the functional group (which is indicative of the resulting leaving group), but also on the characteristics of the substituents; sterics are important to keep in mind when analyzing reactivity.

Things to Watch Out For

Remember that on Test Day you can refer to the Periodic Table; periodic trends, like electronegativity, can help you figure out relative electrophilicity or nucleophilicity.

E Phosphorus-Containing Compounds

Physiologically, phosphoric acid may be found as a neutral compound or as a mono-, di-, or trivalent anion. Accordingly, it has three pK_a values: 2.15, 7.20, and 12.35. What are the relative concentrations of phosphoric acid and its ions at physiological pH (around 7.4)? If a pure sample of phosphoric acid were dissolved in *n*-butanol, would there be a reaction? If so, what would the products be?

1 What is the most abundant form of phosphoric acid, based on pK_a?

The question gives us three pK_a values: 2.15, 7.20, and 12.35. Remember, the concentration of an acid and its conjugate base are equal when pH = pK_a. For example, $[H_3PO_4] = [H_2PO_4^-]$ at pH = 2.15. As the pH increases above 2.15, the concentration of H_3PO_4 will decrease as the concentration of $H_2PO_4^-$ increases. A pH of 7.4 is sufficiently far from 2.15 that there is essentially no H_3PO_4 present at all. Similarly, PO_4^{3-} is also present in negligible quantities at pH 7.4 because pK_{a3} is so high (12.35).

The two intermediate forms, $H_2PO_4^-$ and HPO_4^{2-}, predominate at physiological pH because this pH is close to pK_{a2} for phosphoric acid (7.20). We could predict that because 7.4 is slightly more basic than 7.20, there will be a slightly higher concentration of HPO_4^{2-}, the conjugate base, than $H_2PO_4^-$ at this pH. Indeed, we can determine the relative concentrations from the Henderson–Hasselbalch equation:

$$pH = pK_a + \log\frac{\left[HPO_4^{2-}\right]}{\left[H_2PO_4^-\right]}$$

$$7.4 = 7.20 + \log\frac{\left[HPO_4^{2-}\right]}{\left[H_2PO_4^-\right]}$$

$$0.2 = \log\frac{\left[HPO_4^{2-}\right]}{\left[H_2PO_4^-\right]}$$

$$10^{0.2} = \frac{\left[HPO_4^{2-}\right]}{\left[H_2PO_4^-\right]} \approx 1.58$$

In summary, the relative concentrations of the various conjugates can be expressed as a ratio (H_3PO_4:$H_2PO_4^-$:HPO_4^{2-}:PO_4^{3-}):

$$\sim 0:1:1.58:\sim 0$$

2 What are the possible reaction types for phosphoric acid reacting with *n*-butanol?

This part of the question is asking whether phosphoric acid can react with alcohols. The answer to this question is undoubtedly *yes*, but what is the reaction? Any time an acid is present, we should look for a group to protonate—in this case, the hydroxyl group on *n*-butanol is the best candidate.

Alternatively, phosphoric acid can be viewed as the hydrolyzed form of an organic phosphate. It is reasonable to think that this reaction might be reversible, creating mono-, di-, or tri-*n*-butyl phosphates. Let's examine these two reactions to determine what products can result.

3 What reaction occurs when phosphoric acid protonates *n*-butanol?

As described in Step 2, phosphoric acid could protonate the hydroxyl group of the alcohol. When this happens, the hydroxyl group (which is now a water molecule) becomes a good leaving group, forming a carbocation. While elimination reactions are not tested directly on the MCAT, it is reasonable to predict that the molecule will seek to stabilize or reduce the positive charge; formation of a double bond through an elimination reaction would be one way to accomplish this goal:

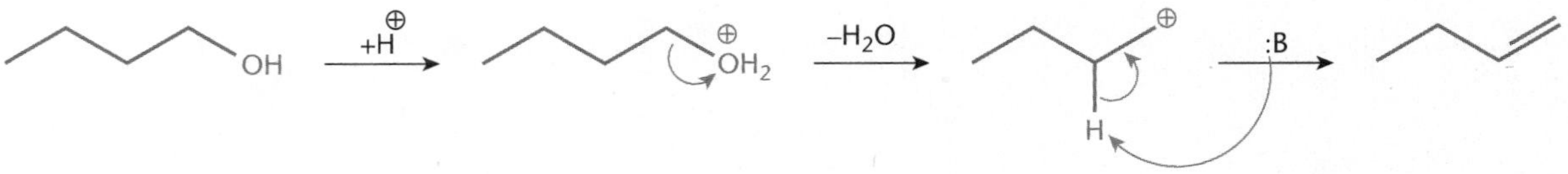

4 What reaction occurs when *n*-butanol attacks phosphoric acid?

Another potential mechanism is the formation of a phosphate ester, which could occur if the butanol attacked the phosphate group. Elimination of a hydroxyl group as water could then follow.

Takeaways

The different forms of phosphoric acid have different pK_a values; phosphoric acid commonly reacts by protonating a leaving group to improve that group's ability to leave.

Things to Watch Out For

When comparing two plausible reactions, consider the relative nucleophilicity, electrophilicity, and leaving group abilities of the reacting compounds.

Note the similarity between this reaction and nucleophilic acyl substitution (specifically, esterification). Given the two additional hydroxyl groups on the phosphorus, one could imagine two more iterations of this mechanism to produce the di- and tributyl esters.

Related Questions

1. Which class of biomolecules *always* contain phosphorus? Which ones *sometimes* contain phosphorus?

2. One way to increase blood pH is to increase the excretion of *titratable acid* (protonated phosphates) from the kidney. Urine is usually slightly more acidic than physiological pH. What would happen to the relative concentrations of the different forms of phosphoric acid at this slightly lower pH?

3. $H_2PO_4^-$ is a Brønsted–Lowry acid because it can give up a proton. Would $H_2PO_4^-$ react in the same fashion with butanol as H_3PO_4 did?

High-Yield Problem-Solving Guide questions continue on the next page. ▶ ▶ ▶

⊟ Spectroscopy

An unknown compound was discovered in an old, unused laboratory. Its molecular formula was determined to be $C_6H_9NO_2$ by high-resolution mass spectrometry. The following IR stretches were recorded: 3300 (sharp), 2890 (medium), 2220, 1740 (sharp), 1220, 984, 700, and 650 cm^{-1}.

The ^1H–NMR spectrum of the compound is as follows:

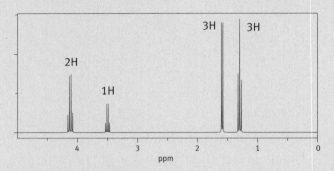

Given this information, determine the structure of the unknown compound.

❶ What functional groups are present based on the IR data?

Here, the most important signal is the sharp peak at 1740 cm^{-1}, which indicates the presence of a carbonyl. A carbonyl indicates that there must be one of the following functional groups in our molecule: aldehyde, ketone, carboxylic acid, anhydride, ester, or amide. It can't be an aldehyde because there are no aldehyde signals in the NMR (a single proton around 9–10), and it can't be a carboxylic acid because there is no hydroxyl group stretch in the IR (a broad absorption around 2800–3200). Nor can it be an anhydride because there are not enough oxygen atoms in the molecular formula—anhydrides require at least three. Because the stretch is close to 1750, it is most likely to be an ester. This is also supported by the ^1H–NMR data, as discussed below.

❷ What is the structure of the esterifying group, according to the ^1H–NMR?

The signal for the esterifying group should be the farthest downfield (to the left) in the ^1H–NMR spectrum because it is attached to the most electronegative element in the molecule. The peak farthest to the left is composed of two hydrogens, which are split into four peaks. Keeping in mind the $n + 1$ rule, this implies that there are two hydrogens on the carbon closest to the ester oxygen and three hydrogens on the next one. This is supported by the three hydrogens split into a triplet around 1.3 ppm. Thus, we know the molecule must contain the following structure:

③ Based on the molecular formula, is the parent chain unsaturated?

So far, we have determined that there is an ester with an ethyl esterifying group. This accounts for three carbon atoms (the carbonyl carbon and two in the esterifying group), two oxygen atoms, and five hydrogen atoms. Subtracting these from the molecular formula, $C_6H_9NO_2$, we are left with three carbon atoms, four hydrogen atoms, and a nitrogen atom. Immediately, we should recognize that there must be at least one multiple bond in the molecule because there are not enough hydrogens to create an alkyl chain. There are three possibilities for the parent chain to get the right numbers: a carbon–carbon triple bond, a carbon–nitrogen triple bond, or two carbon–carbon double bonds.

④ What option for unsaturation is correct, using the ^1H–NMR data?

We can immediately eliminate the two carbon–carbon double bonds as an option because this would mean that at least one carbon is sp^2-hybridized; sp^2-hybridized carbons absorb in the range of 4.6–6, and there are no peaks in this range. Between the carbon–carbon triple bond and the carbon–nitrogen triple bond, then, we can determine that it must be the latter. If there was a carbon–carbon triple bond, the nitrogen would have two hydrogens on it to form an amine, and there are no peaks remaining that consist of only two hydrogens. Therefore, this compound must contain a nitrile.

Takeaways

With these combined structure problems, make sure to utilize all of the data at your disposal. The process is very much like taking the pieces of a jigsaw puzzle and putting them together.

Things to Watch Out For

Be sure to avoid overinterpreting the IR data. Many peaks can correspond to multiple functional groups or are inconclusive. ^{1}H–NMR data, on the other hand, is a wealth of information that, when interpreted carefully, can provide nearly all the information to figure out a molecule's structure.

5 What is the backbone of the parent chain?

If we account for all of the atoms in the ester and nitrile functionalities, we are left with two carbons and four hydrogens. There can only be two possibilities, structurally:

Note that the protons in the structure on the left would have to give rise to two triplets because each is adjacent to a carbon with two protons. However, the only signals we haven't accounted for in the NMR are a doublet integrating for three protons and a quartet integrating for one. These signals exactly match the structure on the right, so that must be the unknown.

Related Questions

1. Why is it unlikely that the compound in the original question is cyclic?

2. If the molecule were a methyl ketone rather than an ester (that is, replacing the ethoxy group with a methyl group), where would you expect that the carbonyl IR stretch would appear? Why?

3. In the compound described in the previous question, where would the signal for the methyl ketone protons show up in the ^{1}H–NMR spectrum, relative to the signal for the protons adjacent to the oxygen in the ester?

High-Yield Problem-Solving Guide questions continue on the next page. ▶ ▶ ▶

Key Concepts

Organic Chemistry Chapter 12
Extraction (liquid–liquid separation)
Acid–base properties
Separation and purification schemes
Polarity

⑤ Solubility-Based Separations

A student studying electrophilic aromatic substitution synthesizes several substituted aromatic compounds from benzene. Following the series of synthesis reactions, the student is left with a mixture of two products, benzoic acid and 4-nitroaniline. However, this mixture is also contaminated with unreacted benzene and unreacted small electrophiles that were used in the synthesis. The student next wishes to separate this mixture. To perform the separation, the liquid mixture is dissolved in 500 mL dichloromethane (density = 1.33 g/mL). The solution is washed with water three times and the aqueous layer (A) is collected. The remaining organic layer is then washed with 20% aqueous NaOH three times, and the aqueous layer (B) is collected. Next, the remaining organic layer is washed with 10% aqueous HCl three times and the aqueous layer (C) is once again collected, leaving behind the organic layer (D). The student discards the two layers containing unreacted contaminants. Finally, the student removes the purified benzoic acid and the purified 4-nitroaniline products from their respective solvents using a rotary evaporator. However, the student observes that the yield of benzoic acid is significantly lower than predicted. What are the contents of layers (A), (B), (C), and (D)?

4-nitroaniline

Benzoic acid

Less Dense Liquid

Charged or polar solutes go to the aqueous layer

Uncharged, nonpolar solutes go to the organic layer

More Dense Liquid

① ### How does extraction use differences in physical properties to separate compounds?

Extraction relies on the principle of "like dissolves like" in order to successfully separate liquid compounds. A separatory funnel is filled with two layers of fluid with different properties. The two liquids often differ in hydrophobicity, but other traits can be used for separation as well. The target substances will be separated in a successful extraction as each compound will only dissolve in one of the two solvents.

② ### How does a separatory funnel facilitate the removal of the small electrophile contaminants into layer (A)?

The mixture is placed in a separatory funnel with dichloromethane, a polar solvent. This polar solvent will dissolve electrophilic molecules, but benzene will be unable to stay in solution with dichloromethane and thus will remain separate from the dichloromethane. When the aqueous layer is drained, the electrophilic contaminants will leave with the dichloromethane.

③ ### Why does aqueous NaOH radically increase benzoic acid's solubility in water?

Sodium hydroxide functions as a strong base, and would deprotonate the relatively weak benzoic acid. This deprotonation would yield benzoate, a negatively charged ion. This charged ion would be able to dissolve easily into water, as compared to the relatively poor solubility of benzoic acid in water.

④ ### How does aqueous HCl facilitate the extraction of 4-nitroaniline into layer (C)?

4-Nitroaniline is a relatively nonpolar compound. On reaction with acid, the nitro group becomes protonated, leaving 4-nitroaniline with a positive charge. This positive charge increases solubility in the aqueous layer and decreases solubility with the nonpolar benzene, allowing 4-nitroanilline to be extracted into the aqueous layer.

Takeaways

In an extraction problem, each compound will either be dissolved in the aqueous layer or the organic layer. However, if a compound is acidic or basic, it is possible to transpose it to the aqueous layer by using basic or acidic washes, respectively.

Things to Watch Out For

Don't assume that the organic phase will be on top—this depends on the densities of the two phases. For example, dichloromethane $\left(1.3 \frac{g}{mL}\right)$ is denser than water $\left(1.0 \frac{g}{mL}\right)$; dichloromethane will sink to the bottom of the separatory funnel, with the water floating on top.

5. What are the contents of each layer?

Based on the extractions as described, the layers would be expected to separate as follows:

Related Questions

1. What would be an appropriate extraction procedure to separate a mixture of phenol and benzoic acid dissolved in ether?

2. In order to extract *p*-nitrophenol from phenol in an ether solution, a student washes the organic layer with 10 mL of a 5% aqueous solution of NaOH. After the washing, what will be left in the organic layer?

3. Why does the student perform three sets of washings in each step of the extraction?

Solutions to Related Questions

1. Nomenclature

1. The alkyl group on the ester oxygen is called the esterifying group. It is named as a separate word before the rest of the compound name. This word should be viewed as an adjective describing the ester; the compound we named was the *ethyl* ester. Therefore, if the ethyl (alkyl) group of ethyl 2-methoxy-3-(2-oxocyclohexyl)propanoate contained substituents, the word *ethyl* would be replaced with whatever group was attached to the ester oxygen, using the suffix –*yl*.

2. Sodium borohydride (NaBH$_4$) is a reducing agent; as such, it will reduce the ketone in this molecule to a secondary alcohol. The two products of the reduction reaction would be (1R,2S)-2,5,5-trimethylcyclohexanol and (1S,2S)-2,5,5-trimethylcyclohexanol, which are shown below:

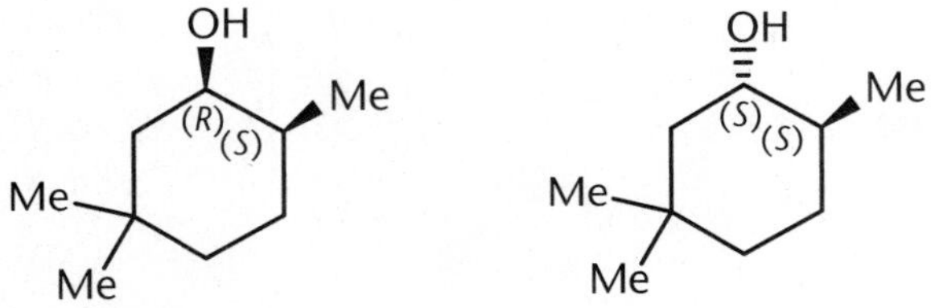

3. Lithium aluminum hydride (LiAlH$_4$) is an even stronger reducing agent than sodium borohydride and will reduce the imine in the molecule to a secondary amine. It is not strong enough to reduce the alkene to an alkane, however. The two products of the indicated reaction would be (S,Z)-6-chloro-6-cyclopentyl-*N*-methylhex-5-en-2-amine and (R,Z)-6-chloro-6-cyclopentyl-*N*-methylhex-5-en-2-amine, which are shown below:

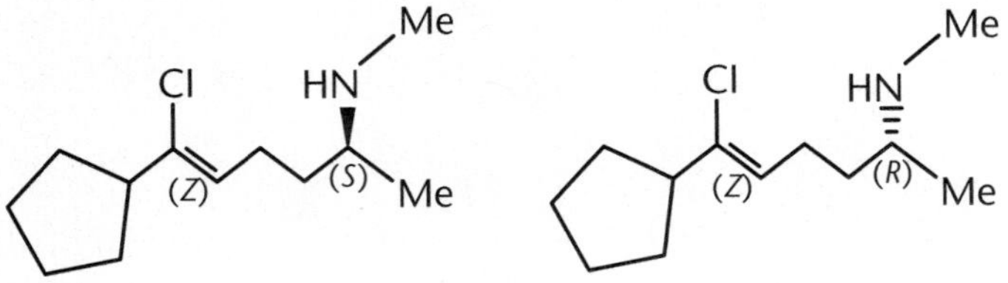

2. Stereoisomers

1. The D- and L-isomers of a compound are enantiomers of each other, which means that they have opposite stereochemistry at all chiral carbons and no internal plane of symmetry. As such, they are nonsuperimposable mirror images. Enantiomers always have opposite specific rotation, so the specific rotation of the L-isomer must be $-140°$. Racemic mixtures contain equal concentrations of both enantiomers. The specific rotations from the two enantiomers cancel each other, resulting in an optically inactive solution ($0°$ rotation).

2. Only the middle compound would give rise to an achiral polyol upon reduction. The other two would not contain a plane of symmetry.

3. For the purposes of the MCAT, it is not possible to have a chiral compound that does not contain chiral centers. Chiral compounds not only need to contain chiral centers, but must also have the right geometry, such that the chiral centers do not cancel each other out (as occurs in a *meso* compound). There are rare examples of complex polyaromatic compounds and certain allenes that do show chirality despite a lack of chiral centers, but such compounds are outside the scope of the MCAT.

3. Hybridized Orbitals

1. The lower the pK_a of a compound, the more acidic it is. Similarly, the lower the pK_b, the more basic it is. Compounds with low pK_b values are stronger bases. As such, they are not good proton donors (Brønsted–Lowry acids), preferring to accept protons (Brønsted–Lowry bases). Therefore, the higher the pK_b, the more acidic (less basic) the compound.

2. Carbon dioxide contains two double bonds. This requires two unhybridized p-orbitals—one to participate in each of the π bonds. Therefore, the hybridization of the carbon must be sp. Be careful not to fall into the trap of assuming that the presence of double bonds automatically means sp^2 hybridization. Indeed, because there are only two areas of electron density in carbon dioxide, the carbon must be sp-hybridized.

3. Careful reading of this compound's name indicates that there is only one double bond in the molecule, implied by the suffix *-one*. If we draw out the structure of 5-hydroxy-2,6-dimethyl-3-heptanone, it confirms that the highest bond order is 2 (double bond):

4. Nucleophilicity Trends

1. Because all of the nucleophiles contain the same attacking atom (nitrogen), nucleophilicity will parallel basicity, irrespective of the solvent used. Here, resonance delocalization and hybridization each play a role.

First, we will compare pyridine, triethylamine, and acetonitrile, as they each have only one nitrogen atom with one isolated lone pair. It may appear that the lone pair on the pyridine nitrogen may be resonance-stabilized, but in fact that lone pair does not engage in the aromatic ring delocalization. The nitrogen in acetonitrile is *sp*-hybridized, the nitrogen in pyridine is *sp²*-hybridized, and the nitrogen in triethylamine is *sp³*-hybridized. In a similar manner to the carbon atoms discussed in the *Hybridized Orbitals* question, the *sp*-hybridized lone pair will be the most stable (least basic), while the least stable (most basic) is the pair on triethylamine. The basicity order for these three compounds is acetonitrile < pyridine < triethylamine.

Now we must consider DMAP. The heterocyclic nitrogen (the nitrogen in the ring) is similar to pyridine, while the other nitrogen looks similar to that of triethylamine. However, while the lone pair of the ring-bound nitrogen is not delocalized (similar to pyridine), the lone pair of the tertiary nitrogen does form resonance structures with the ring. This resonance delocalization makes those electrons much less basic, and so the heterocyclic nitrogen is the one that acts as the base. Now we must decide whether DMAP is more or less basic than pyridine. When DMAP gains a proton, the positive charge is stabilized via resonance with the amino substituent, making DMAP more likely to gain a proton than pyridine, thereby making it more basic (and nucleophilic).

The final order is acetonitrile < pyridine < DMAP < triethylamine.

2. Methanol is a polar protic solvent, so we should be careful when considering how nucleophilicity and basicity relate. The halogens are all in the same group, so the stronger the base, the *weaker* the nucleophile in a polar protic solvent. To assess the basicity of the halogens, consider the strength of their conjugate acids, HF, HCl, HBr, and HI. The strongest acid (HI) has the weakest conjugate base, and the weakest acid (HF) has the strongest conjugate base. Therefore, the nucleophilicity for the halogens in a polar protic solvent like methanol is $I^- > Br^- > Cl^- > F^-$.

DMSO is a polar aprotic solvent. Nucleophilicity always parallels basicity in polar aprotic solvents, so the nucleophilicity trend in DMSO is $F^- > Cl^- > Br^- > I^-$.

3. The order of increasing nucleophilicity in methanol is $Ph_3N < Et_3N < Ph_3P < Et_3P < Et_3As$. Triphenylamine and triphenylphosphine are less nucleophilic than triethylamine and triethylphosphine, respectively, because in the phenyl analogs, the lone pair can be delocalized into the three phenyl rings, making it less reactive and less nucleophilic. As for comparing the nitrogen, phosphorus, and arsenic compounds, the amine is more basic than the phosphine, which is more basic than the arsine. Because these are all in the same group and the solvent (methanol) is protic, the nucleophilicity is backwards from basicity, making arsenic the most nucleophilic of the group.

5. Reactions of Alcohols

1. Changing the hydroxyl group to a stronger nucleophile would change the mechanism by which it proceeds to S_N2 (or E2). S_N2 reactions proceed via a one-step (concerted) mechanism. While S_N1 and S_N2 reactions often have pronounced differences in stereochemistry, in this case the stereochemistry would be the same. This is because the nucleophile in the first step will have to attack from above the ring, resulting in an epoxide (or other three-membered ring intermediate) sticking out of the plane of the page, regardless of the nucleophilic substitution mechanism used. Thus, while the mechanism would be different with a stronger nucleophile, the products would be the same.

2. Changing the tosylate group to a hydroxyl would leave us without a good leaving group. While this reaction could still occur under acidic conditions (because acid would protonate the hydroxyl groups, making them better leaving groups), it would no longer occur under the conditions given.

3. In this case, it is not possible to favor one product over the other without changing the reactant molecule. The two compounds are almost identical in stability, meaning that both will form in approximately equal amounts. If the reactant molecule could be altered by changing one of the methyl substituents, or by adding another substituent to the ring, this could sterically impede the ring-opening nucleophile and favor one product over the other.

6. Identifying the Structure of an Unknown Oxy Compound

1. Product **E** is one of four possible stereoisomers. While **D** and **E** are diastereomers of each other, each also has an enantiomer that has opposite absolute configurations at both the nitrogen-bearing carbon and the methyl group on C-3. Looking just at the stereocenters, we could describe these compounds as: ($1R,3R$), ($1R,3S$), ($1S,3R$), and ($1S,3S$).

2. The products in this reaction are diastereomers of each other. While a racemic mixture (equimolar concentrations of two enantiomers) should have no optical activity, this is because enantiomers always have specific rotations with equal magnitude but opposite sign—canceling each other in solution. Diastereomers do not have the same predictable relationship; knowing the specific rotation of one diastereomer gives no indication of the specific rotation of another. Therefore, we would not expect the specific rotations of the two diastereomers to cancel, and a solution containing equimolar concentrations of two diastereomers *should* have optical activity.

3. *Cis/trans* isomers differ by the positions of groups around an immovable bond. This can refer to a double bond (C=C, C=N, C=O), but can also refer to the bonds in a cyclic molecule. Cycloalkanes can also have *cis/trans* isomers.

7. Enolate Chemistry

1. The most acidic atoms in cyclohexanone are the α-carbons, which readily give up a proton in the presence of a base. This results in the formation of a secondary carbanion, which can push its lone pair to form a double bond with the carbonyl carbon. This converts the carbonyl to the enolate intermediate.

2. Conjugation refers to alternating single and multiple bonds, which necessarily means that there are unhybridized p-orbitals aligned with each other. This allows delocalization of π electrons throughout those p-orbitals, resulting in a cloud of electron density above and below the plane of the molecule. Delocalization of these electrons stabilizes a molecule because electrons are able to spread out over a larger volume of space, decreasing their energy. Conjugated molecules can stabilize charges, as well, by delocalizing them throughout the π-electron clouds.

3. 2,2,6,6-Tetramethylcyclohexanone does not have any α-hydrogens because both α-carbons are quaternary carbons, as shown below:

4,4-Dimethylhexanal, on the other hand, has α-hydrogens on its α-carbon because the α-carbon is secondary, as shown below:

Because 2,2,6,6-tetramethylcyclohexanone does not contain any α-hydrogens, it cannot be deprotonated at the α-carbon to form the nucleophilic enolate. Thus, it must act as the electrophile. 4,4-Dimethylhexanal will form the enolate and will serve as the nucleophile in this reaction.

8. Nucleophilic Acyl Substitution

1. Methanol is generally a better leaving group than ethanol. Alkyl chains are electron-donating and therefore destabilize the negative charge that remains on the alkoxide oxygen after heterolysis. (They also destabilize the lone pairs that remain on the hydroxyl oxygen, under acidic conditions.) This effect is less pronounced for the shorter alkyl chain of methanol than for the longer alkyl chain of ethanol.

2. Lactones are cyclic esters. They can be synthesized by intramolecular attack in a molecule that contains both a hydroxyl group and a carboxylic acid. The nucleophilic hydroxyl group attacks the carbonyl carbon, as shown below:

3. Lactams are cyclic amides. They can be synthesized by intramolecular attack in a molecule that contains both an amine and a carboxylic acid, in a similar mechanism to lactone formation. An example of a lactam is shown below:

9. Properties of Carboxylic Acid Derivatives

1. Peptide bonds join amino acids together in proteins. Many proteins in the body serve a structural role or must withstand tension, both of which require that the integrity of the amino acid sequence be maintained. The body is also an aqueous environment. Under normal conditions, water does not react with an amide to form a carboxylic acid and an amine (hydrolysis). Amide bonds are stable and relatively nonreactive; were they more reactive than carboxylic acids, life as we know it would not exist.

2. Aldehydes are always terminal groups and, as such, have less steric hindrance than comparable ketones, which are always internal in a compound. By extension, an aldehyde will be more reactive toward a nucleophile.

3. Hydrogen bonding occurs in molecules containing a hydrogen bound to a very electronegative atom (N, O, or F). Amides may have hydrogens on the amide nitrogen and can therefore hydrogen bond. Carboxylic acids can also hydrogen bond because they have a hydrogen on the acyl oxygen. Under normal circumstances, neither anhydrides nor esters can hydrogen bond.

10. Phosphorus-Containing Compounds

1. Out of the four major classes of biomolecules (carbohydrates, lipids, proteins, and nucleic acids), only nucleic acids *require* phosphorus. It is a component of adenosine triphosphate (ATP), which is used for energy storage, and also of organic phosphates, like the nucleotides that make up RNA and DNA. On the other hand, phosphate groups can be found attached to at least *some* lipids, proteins, and carbohydrates (for example, the phospholipids in the cell membrane).

2. As the pH drops slightly from physiological pH, the concentration of $H_2PO_4^-$ will increase and the concentration of HPO_4^{2-} will decrease; lower pH means more protons are available in solution, so the protonated form will become more favorable. At a pH of 7.20 (pK_{a2} of phosphoric acid), the concentrations of these two species will be equal. At a pH of 7, the ratio between the two species will be reversed from the ratio at 7.4. That is, $H_2PO_4^- : HPO_4^{2-} = 1.58 : 1$. Note that the concentrations of phosphoric acid and phosphate anion will remain essentially zero; the pH is still far away from the pK_a.

3. $H_2PO_4^-$ is not nearly as strong of an acid as H_3PO_4—pK_{a2} is over five pH scale values higher than pK_{a1}! As a much weaker acid, $H_2PO_4^-$ cannot protonate the alcohol to any significant extent, so the dehydration reaction will not occur.

11. Spectroscopy

1. There are only six carbon atoms in the molecular formula, which means that any rings present would have to be small. Small rings have a high degree of angle strain (especially if there are double or triple bonds in the ring), which would make having one or multiple rings in the unknown very unlikely.

2. If the molecule were a methyl ketone rather than an ester, the carbonyl IR stretch would be closer to 1700 cm^{-1}. The resonance donation of the ester alkoxide makes the C=O bond more unstable and therefore higher in energy.

3. The methyl ketone protons would be further upfield in the ^{1}H–NMR spectrum (closer to 2 ppm than 4 ppm). This is because the alkyl group in the ketone is less electron-withdrawing than the highly electronegative oxygen in the ester.

12. Solubility-Based Separations

1. To separate a mixture of benzoic acid and phenol dissolved in ether, it would be best to take advantage of their relative acidities. Benzoic acid is a stronger acid than phenol, so a weak base ($NaHCO_3$, for example) can be used to deprotonate benzoic acid and move it into the aqueous phase, while leaving phenol in the organic phase.

2. After washing the mixture with 5% NaOH, p-nitrophenol and phenol will both be deprotonated by the strong base and will move into the aqueous phase. Nothing will be left in the organic layer except for the ether solvent and water-insoluble impurities.

3. Each compound has a given solubility in water, which may be exceeded if it has a high concentration in the organic solvent. Further, not all of the desired product necessarily comes into contact with the water during the first extraction and may remain dissolved in the organic solvent. In general, it is more effective to perform multiple small washes, as done by this student, than to try to extract all of a desired product in one step with a larger amount of solvent.

Physics and Math

E Inclined Plane

A block with a mass of 2 kg is sliding down a plane that is inclined at 30° to the horizontal. The coefficient of kinetic friction between the block and the plane is 0.3. Starting from rest, how far does the block travel in 2 seconds?

1 **What free body diagram would you generate for the forces present?**

There are three forces acting on the block: the force of gravity ($\mathbf{F}_g$, the magnitude of which is mg), which always acts straight down; kinetic friction (labeled $\mathbf{f}_k$), which always acts opposite the direction of motion; and the normal force (labeled $\mathbf{N}$), which is always perpendicular to the plane:

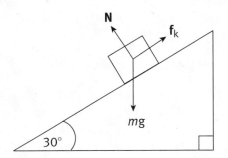

2 **How would you adjust for the parallel and perpendicular components of weight?**

Orient the axes in this question to be parallel and perpendicular to the plane; this will make calculations much easier because we can solve for the acceleration in the parallel direction. $\mathbf{N}$ points in the positive perpendicular direction, and $\mathbf{f}_k$ points in the negative parallel direction. The weight, $\mathbf{F}_g$, must be broken into components along these axes:

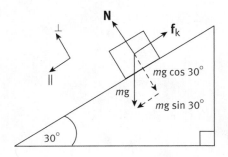

The parallel component of the gravitational force, $\mathbf{F}_{g,\parallel}$, is equal to $mg \sin \theta$. The perpendicular component of the gravitational force, $\mathbf{F}_{g,\perp}$, is equal to $mg \cos \theta$.

3 What is the sum of forces in each direction?

The sum of the forces in a given direction is always equal to $m\mathbf{a}$ according to Newton's second law:

$$\mathbf{F}_{net,\parallel} = m\mathbf{a}_{\parallel} = mg \sin 30° - \mathbf{f}_k$$

$$\mathbf{F}_{net,\perp} = m\mathbf{a}_{\perp} = \mathbf{N} - mg \cos 30°$$

4 What is the normal force?

We know that the block is not accelerating in the perpendicular direction because it is not sinking into the plane or coming off of the plane; thus, we can set $\mathbf{a}_{\perp} = 0$ and solve for $\mathbf{N}$:

$$\mathbf{a}_{\perp} = 0$$

$$\mathbf{N} - mg \cos 30° = 0 \rightarrow \mathbf{N} = mg \cos 30°$$

5 What is the acceleration of the block?

The force of friction depends on the magnitude of the normal force, N, and the coefficient of kinetic friction, μ_k. Plug in the expression for normal force from Step 4 to determine the frictional force:

$$f_k = \mu_k N = \mu_k mg \cos 30°$$

Then, plug this expression for frictional force into the forces in the parallel direction expression from Step 3 to determine the acceleration:

$$m\mathbf{a}_{\parallel} = mg \sin 30° - \mathbf{f}_k$$

$$m\mathbf{a}_{\parallel} = mg \sin 30° - \mu_k mg \cos 30°$$

$$\mathbf{a}_{\parallel} = g(\sin 30° - \mu_k \cos 30°)$$

$$\approx \left(10 \, \frac{m}{s^2}\right)\left(\frac{1}{2} - 0.3 \times 0.866\right)$$

$$\approx \left(10 \, \frac{m}{s^2}\right)\left(\frac{1}{2} - 0.27\right) = 10 \times 0.23 = 2.3 \, \frac{m}{s^2}\left(actual = 2.40 \, \frac{m}{s^2}\right)$$

Note that the mass cancels out completely.

Takeaways

In every force problem, the process is the same: draw a free body diagram, write Newton's second law expressions for each dimension, and then solve. In most single-entity problems, the mass will cancel out of the equation. Notice that this problem would be much simpler if the ramp were frictionless.

Things to Watch Out For

The geometry can become confusing on Test Day, so it is helpful to memorize that the component of the gravitational force along the plane is $mg \sin \theta$, where θ is the angle of the ramp relative to the horizontal.

6 What is the displacement?

We are given an acceleration, an initial velocity (starting from rest), and a time. Thus, we can determine the displacement:

$$\Delta \mathbf{x} = \mathbf{v}_0 t + \frac{\mathbf{a}t^2}{2}$$

$$= 0 + \frac{\left(2.4 \, \frac{\mathrm{m}}{\mathrm{s}^2}\right)\left(2 \, \mathrm{s}\right)^2}{2} = 4.8 \, \mathrm{m}$$

Related Questions

1. A block of mass 5 kg is placed on an inclined plane at 45° to the horizontal. What is the minimum coefficient of static friction so that the block remains at rest?

2. A block is given an initial velocity of $2 \, \frac{\mathrm{m}}{\mathrm{s}}$ up a frictionless plane inclined at 60° to the horizontal. What is the highest point above the original height reached by the block?

3. What is the velocity of a 10 kg block sliding down a frictionless inclined plane at 30° to the horizontal 5 seconds after it is released from rest?

High-Yield Problem-Solving Guide questions continue on the next page. ▶ ▶ ▶

E Projectile Motion and Air Resistance

An arrow with a mass of 80 g is fired at an angle of 30° to the horizontal. It strikes a target located 5 m above the firing point and impacts the target traveling at a speed of 20 $\frac{\text{m}}{\text{s}}$. If 10 percent of the initial energy of the arrow is lost to air resistance, what was the initial speed of the arrow?

1 What would be the equation for the final energy of the arrow?

The total energy of the arrow is its potential energy plus its kinetic energy. The potential energy of the arrow is *mgh*. For simplicity, make the height at the firing point the datum ($h = 0$) so that the final height is 5 m. Remember that finding a numerical value at this point is not necessary. Writing a variable expression will save you time because some term (usually the mass) may cancel out in a later step. Thus, the final energy is:

$$E_f = K_f + U_f = \frac{1}{2}mv_f^2 + mgh_f$$

2 What would be the equation for the initial energy of the arrow?

As stated in Step 1, the initial height is assigned the value of zero, so the potential energy at this point is also zero:

$$E_i = K_i + U_i = \frac{1}{2}mv_i^2 + mgh_i = \frac{1}{2}mv_i^2$$

3 How can you relate the change in energy to the energy lost to air resistance?

The conservation of energy equation tells us that all of the energy of a system must be accounted for. Whatever energy is lost between the beginning and the end must have been due to air resistance:

$$W_{\text{nonconservative}} = E_i - E_f$$
$$= \frac{1}{2}mv_i^2 - \left(\frac{1}{2}mv_f^2 + mgh_f\right)$$

4 ## What is the relationship between the energy lost and initial energy?

The question states that 10 percent of the initial energy is lost. Thus, the energy lost is the initial energy times 0.1:

$$W_{\text{nonconservative}} = 0.1 \times E_i = 0.1\left(\frac{1}{2} \times mv_i^2\right)$$

5 ## What is the initial speed?

$$W_{\text{nonconservative}} = \frac{1}{2}mv_i^2 - \left(\frac{1}{2}mv_f^2 + mgh_f\right)$$

$$0.1 \times v_i^2 = v_i^2 - \left(v_f^2 + 2gh_f\right)$$

$$0.9 \times v_i^2 = v_f^2 + 2gh_f$$

$$v_i = \sqrt{\frac{v_f^2 + 2gh_f}{0.9}} \approx \sqrt{\frac{\left(20\,\frac{m}{s}\right)^2 + 2 \times 10\,\frac{m}{s^2} \times 5\,m}{0.9}}$$

$$= \sqrt{\frac{400 + 100}{0.9}} \approx \sqrt{529} = 23\,\frac{m}{s}\left(\text{actual} = 23.5\,\frac{m}{s}\right)$$

Related Questions

1. A rock is dropped from the top of a 100 m tall building and lands while traveling at a speed of $30\,\frac{m}{s}$. How much energy was lost due to air resistance?

2. Two different objects are dropped from rest off of a 50 m tall cliff. One lands going 30 percent faster than the other. The two objects have the same mass. How much more kinetic energy does one object have at the landing than the other?

3. A projectile is fired vertically at a speed of $30.0\,\frac{m}{s}$. It reaches a maximum height of 44.1 m. What fraction of its initial energy has been lost to air resistance at this point?

☰ Thermodynamics

A theoretical ideal engine, known as a Carnot engine, uses the thermodynamic properties of a closed gas piston system. The cycle involves two isothermal steps—one of expansion and one of compression—as well as an expansive and a compressive adiabatic step. This four-step system is shown in a pressure–volume diagram below:

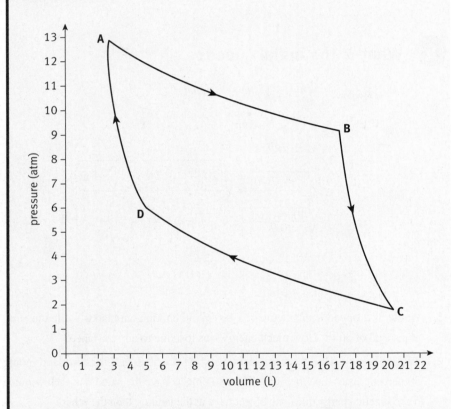

What is the total heat, work, and internal energy change over the course of a single cycle beginning at point **A** and moving through points **B**, **C**, and **D**?

1 What are the processes involved, and what are their properties?

First, we must identify the individual steps as well as the entire process as known thermodynamic functions. Both isothermal and adiabatic processes appear hyperbolic on a pressure–volume graph, but adiabatic processes are steeper than isothermal ones. The path from **A** to **B** is not very steep and thus is an isothermal process; because the volume increases, it is an expansion. The path from **B** to **C** is steeper and is an adiabatic expansion. The cycle then proceeds through an isothermal compression and an adiabatic compression to return to **A**.

In addition to the four steps of this process, we must note that it is a closed system as stated in the question stem; thus, only energy may be exchanged between the system and the environment. It is also a closed cycle, which means that the internal energy at the start and finish must be the same.

② What are the known mathematical relationships?

For thermodynamic relationships, we have certain known quantities. The first law of thermodynamics can be stated as $\Delta U = Q - W$, where ΔU is the change in internal energy, Q is the heat entering the system, and W is the work done by the system.

For the closed cycle from **A** to **B** to **C** to **D** and back to **A**, $\Delta U = 0 = Q - W$; thus, $Q = W$. We need only determine one quantity to determine the second as well. For the isothermal processes between **A** and **B** and between **C** and **D**, the same relationship is true. Because temperature remains constant, internal energy also remains constant and $\Delta U = 0 = Q - W$; thus, $Q = W$.

For the adiabatic processes between **B** and **C** and between **D** and **A**, there is no heat exchange with the environment. The first law can simplify to $\Delta U = 0 - W$; thus, $\Delta U = -W$.

Finally, we know that work can be calculated from the area under the curve in a pressure–volume graph; thus, we can obtain work from the graph by approximating the area using trapezoids.

③ What is the work from each process, and what is the total work?

Work is determined graphically as the area under a pressure–volume curve. For Test Day, it is useful to approximate the area under the curve using trapezoids. The area of a trapezoid is given by:

$$A_{\text{trapezoid}} = \left(\frac{b_1 + b_2}{2}\right)h$$

Keep in mind that the bases b_1 and b_2 are vertical in this graph and the height h is the distance along the x-axis. For the isothermal expansion from **A** to **B**, the calculation of the area under the curve is modeled here:

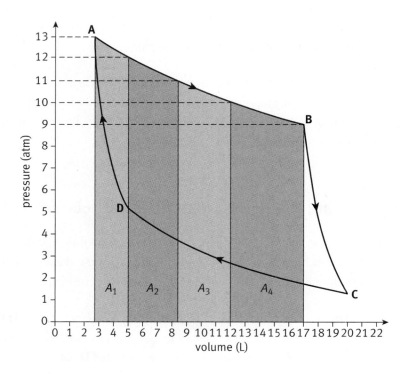

Work is done by the gas as it expands. This is considered positive work. Heat is also added to the gas. To calculate the area, take the average of the bases and multiply by the height:

$$A_1 = \frac{13\,\text{atm} + 12\,\text{atm}}{2}\left(5 - 2.5\,\text{L}\right) = 31.25\,\text{L} \cdot \text{atm}$$

$$A_2 = \frac{12\,\text{atm} + 11\,\text{atm}}{2}\left(8.5 - 5\,\text{L}\right) = 40.25\,\text{L} \cdot \text{atm}$$

$$A_3 = \frac{11\,\text{atm} + 10\,\text{atm}}{2}\left(12 - 8.5\,\text{L}\right) = 36.75\,\text{L} \cdot \text{atm}$$

$$A_4 = \frac{10\,\text{atm} + 9\,\text{atm}}{2}\left(17 - 12\,\text{L}\right) = 47.5\,\text{L} \cdot \text{atm}$$

The total area is $A_1 + A_2 + A_3 + A_4 = 31.25 + 40.25 + 36.75 + 47.5 = 155.75\ \text{L} \cdot \text{atm}$.

This is an unusual unit for work, and even more unusual for heat. It makes sense to convert this answer to joules. Using the conversion factor 1 atm = 101,325 Pa, we can obtain the work in joules:

$$1 \text{ atm} \approx 10^5 \text{ Pa}$$

$$1 \text{ L} \cdot \text{atm} = 10^5 \text{ L} \cdot \text{Pa}$$

$$= 10^5 \text{ L} \cdot \text{Pa} \times \left[\frac{1000 \text{ mL}}{1 \text{ L}} \right] \times \left[\frac{1 \text{ cm}^3}{1 \text{ mL}} \right] \times \left[\frac{1 \text{ m}}{100 \text{ cm}} \right]^3 \times \left[\frac{1 \frac{\text{N}}{\text{m}^2}}{1 \text{ Pa}} \right]$$

$$= \frac{10^5 \times 10^3}{10^6} = 100 \text{ J}$$

Thus, $155.75 \text{ L} \cdot \text{atm} \approx 15.5 \text{ kJ}$ (actual $= 15.8 \text{ kJ}$).

This process must be repeated for the path from **B** to **C**. The same method is used for the path from **C** to **D** and from **D** to **A**, but the values calculated are negative because the gas is undergoing compression. The total work is the sum of the work for these four paths:

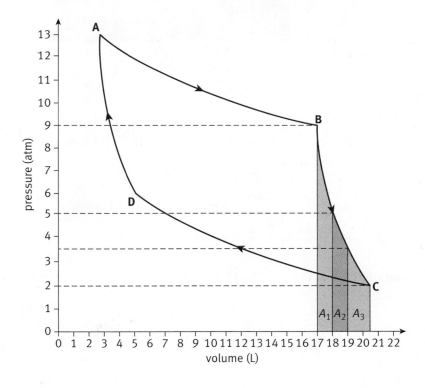

$$A_1 = \frac{9 \text{ atm} + 5 \text{ atm}}{2} (18 - 17 \text{ L}) = 7 \text{ L} \cdot \text{atm}$$

$$A_2 = \frac{5 \text{ atm} + 3.5 \text{ atm}}{2} (19 - 18 \text{ L}) = 4.25 \text{ L} \cdot \text{atm}$$

$$A_3 = \frac{3.5 \text{ atm} + 2 \text{ atm}}{2} (20.5 - 19 \text{ L}) = 4.125 \text{ L} \cdot \text{atm}$$

The total area for **B** to **C** is $7 + 4.25 + 4.125 = 15.375 \text{ L} \cdot \text{atm} \approx 1.5 \text{ kJ}$ (actual $= 1.56 \text{ kJ}$).

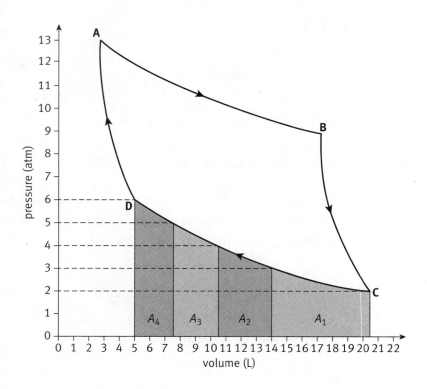

$$A_1 = \frac{2\,\text{atm} + 3\,\text{atm}}{2}\left(14 - 20.5\,\text{L}\right) = -16.25\ \text{L} \cdot \text{atm}$$

$$A_2 = \frac{3\,\text{atm} + 4\,\text{atm}}{2}\left(10.5 - 14\,\text{L}\right) = -12.25\ \text{L} \cdot \text{atm}$$

$$A_3 = \frac{4\,\text{atm} + 5\,\text{atm}}{2}\left(7.5 - 10.5\,\text{L}\right) = -13.5\ \text{L} \cdot \text{atm}$$

$$A_4 = \frac{5\,\text{atm} + 6\,\text{atm}}{2}\left(5 - 7.5\,\text{L}\right) = -13.75\ \text{L} \cdot \text{atm}$$

The total area for **C** to **D** is $-16.25 - 12.25 - 13.5 - 13.75 = -55.75$ L • atm ≈ -5.6 kJ (actual = -5.65 kJ).

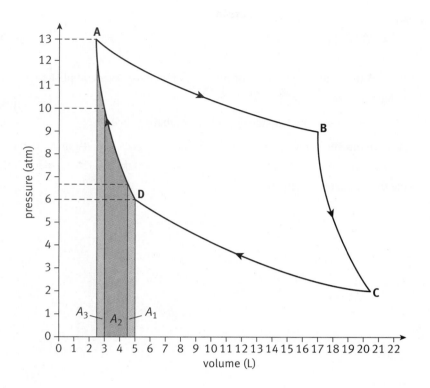

$$A_1 = \frac{6\,\text{atm} + 6.5\,\text{atm}}{2}\left(4.5 - 5\,\text{L}\right) = -3.125 \text{ L} \bullet \text{atm}$$

$$A_2 = \frac{6.5\,\text{atm} + 10\,\text{atm}}{2}\left(3 - 4.5\,\text{L}\right) = -12.375 \text{ L} \bullet \text{atm}$$

$$A_3 = \frac{10\,\text{atm} + 13\,\text{atm}}{2}\left(2.5 - 3\,\text{L}\right) = -5.75 \text{ L} \bullet \text{atm}$$

The total area for **D** to **A** is $-3.125 - 12.375 - 5.75 = -21.25$ L $\bullet$ atm ≈ -2.1 kJ (actual $= -2.15$ kJ).

The total work throughout the cycle is the sum of the work for each portion of the cycle:

$$15.8 + 1.56 - 5.65 - 2.15 = 9.54 \text{ kJ}$$

4. What is the heat absorption during this process?

As established earlier, for a closed-loop process, $\Delta U = 0 = Q - W$; thus, $Q = W$. If the system has performed 9.54 kJ of work, then it has also absorbed 9.54 kJ of heat during the process.

Takeaways

Sign conventions play a significant role in thermodynamic problems. $Q > 0$ means that the system absorbs heat (endothermic), whereas $Q < 0$ means that the system gives off heat (exothermic). $W > 0$ means that the system performs work (expansion), whereas $W < 0$ means that work is done on the system (compression). $\Delta U > 0$ means the temperature increases, whereas $\Delta U < 0$ means the temperature decreases.

Things to Watch Out For

Isothermal processes are often mistaken for adiabatic processes and vice versa. Remember that in an isothermal process, heat is exchanged for work, so heat must enter or leave the system to maintain a constant temperature.

Related Questions

1. What is the minimum number of processes that must be accomplished in order to have a closed cycle?

2. How does the entropy change during an adiabatic process?

3. A thermometer immersed in very hot water undergoes expansion as it changes temperature. Describe the thermodynamic processes that are taking place within the thermometer and the water.

High-Yield Problem-Solving Guide questions continue on the next page. ▶ ▶ ▶

S Fluid Dynamics

A water storage tank is located 300 m away from a water outlet, as shown in the diagram. The top of the tank is vented so that the pressure inside the tank is 1 atmosphere (~1 × 10⁵ Pa). The storage tank has a diameter of 5 m, and the outlet has a diameter of 1 cm. What is the speed of the water exiting the outlet?

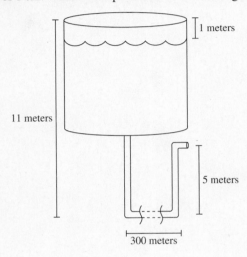

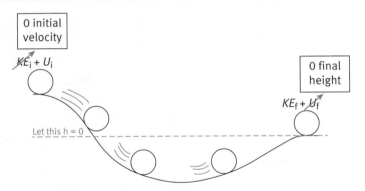

1 In what ways is this situation similar to a rock rolling down a hill?

Both the water storage tank and a rock rolling down a hill would be expected to demonstrate conservation of mechanical energy. In the case of the rock rolling down the hill, as the rock loses gravitational potential energy, it gains an equal amount of kinetic energy (*KE*). Initially, the rock is not moving and has no *KE*. At the end of its path, if you define final height as zero, there is no potential energy. This same logic can be applied to the water storage tank. The top of the tank is analogous to the rock at the top of the hill, and the outlet is analogous to the rock at the bottom of the hill. Potential energy at the top must be equal to *KE* at the bottom.

2 How does Bernoulli's equation express conservation of energy for fluids?

Typically, you would define kinetic energy and potential energy as:

$$KE = \frac{1}{2}mv^2; PE = mgh$$

These expressions are also seen in Bernoulli's equation, albeit in a modified format:

$$P_i + \frac{1}{2}\rho v_i^2 + \rho g h_i = P_f + \frac{1}{2}\rho v_f^2 + \rho g h_f$$

In Bernoulli's equation, the second term $(1/2\rho v^2)$ and third term (ρgh) on each side of the equilibrium equation can be viewed as measures of kinetic and potential energy (PE), respectively, and the mass variable has been converted to density. Density is a measure of mass per unit volume. Thus, these two terms literally represent kinetic energy per unit volume and potential energy per unit volume. The equation dictates that initial KE (per unit volume), PE (per unit volume), and pressure must be equal to final PE (per unit volume), KE (per unit volume), and pressure, which is a demonstration of the conservation of energy.

3 How can the pipes below the tank be simplified to eliminate the final potential energy variable?

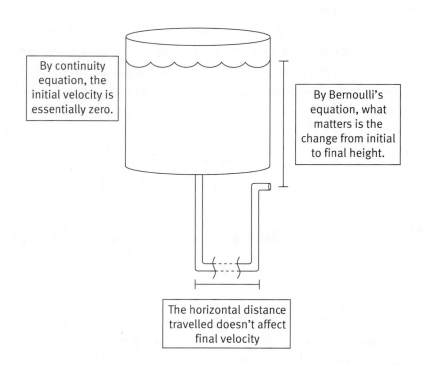

By continuity equation, the initial velocity is essentially zero.

By Bernoulli's equation, what matters is the change from initial to final height.

The horizontal distance travelled doesn't affect final velocity

Takeaways

Bernoulli's equation looks complicated, but it is really just a statement of the conservation of energy. The process is the same for every problem:
1. Write Bernoulli's equation at the two points of interest.
2. Eliminate any variables if possible (often via the continuity equation).
3. Solve for the unknown quantity.

Things to Watch Out For

A common use of Bernoulli's equation is with no change in height, so that $P + \frac{1}{2}\rho v^2 = $ constant. In this situation, a decrease in pressure is associated with an increase in velocity. This is known as the Bernoulli effect and is responsible for balls curving in flight, windows exploding during hurricanes, and (partially) for airplane wings experiencing lift.

For calculation purposes, the initial and final height can be set to any value, provided they maintain the same relationship with one another. As with the ball rolling down the hill, the final height of the pipe can be set to zero, as height in between initial and final points does not affect calculations of conservation of energy. Likewise, horizontal distance travelled and all other measurements that include the path taken do not affect conservative systems. So, recognizing that the horizontal component doesn't matter, and choosing the final pipe height as zero simplifies the equation for this system:

$$P_i + \frac{1}{2}\rho v_i^2 + \rho g h_i = P_f + \frac{1}{2}\rho v_f^2$$

The new initial height must be equal to the difference between the original initial height and the original final height:

$$h_i = (11-1) \text{ m} - 5 \text{ m} = 5 \text{ m}$$

4 How does the continuity equation help eliminate the initial velocity variable?

The continuity equation relates linear flow speed to area:

$$A_i v_i = A_f v_f$$

This equation shows that the product of the cross-sectional area and linear flow speed is a constant. Given that the diameter of the water tank is 5 m and the spout is 1 cm, the linear flow speed at the surface of the tank is near zero. Think about it like this: as the vessel drains, the surface will drop at a nearly negligible speed compared to the speed of the fluid leaving the spout. This approximation further simplifies the equation:

$$P_i + \rho g h_i = P_f + \frac{1}{2}\rho v_f^2$$

5 What is the fluid's velocity as it exits the pipe?

$$P_i + \rho g h_i = P_f + \frac{1}{2}\rho v_f^2$$

We are told that initial and final atmospheric pressures are the same, meaning that pressure can also be removed from the calculation as the values will cancel. This final observation allows the final equation to be in simplified form:

$$\rho g h_i = \frac{1}{2}\rho v_f^2$$

$$\frac{1000 \text{ kg}}{\text{m}^3} \times \frac{10 \text{ m}}{\text{s}^2} \times 5 \text{ m} = \frac{1}{2} \times \frac{1000 \text{ kg}}{\text{m}^3} \times v_f^2$$

$$\frac{5000 \text{ kg}}{\text{m s}^2} = \frac{500 \text{ kg}}{\text{m}^3} \times v_f^2$$

$$100 \frac{\text{m}^2}{\text{s}^2} = v_f^2$$

$$10 \text{ m/s} = v_f$$

Related Questions

1. The pressure at one point in a horizontal pipe is triple the pressure at another point. How do the linear flow speeds compare at these two points?

2. A tapered pipe has two points labeled. At point A, the radius is twice that at point B. The pipe is horizontal and is subjected to a fluid pressure of 1.6 atm. What is the ratio of linear flow speeds of these two points in the pipe?

3. A water storage tank is open to air on the top and has a height of 1 m. If the tank is completely full and a hole is made at the center of the wall of the tank, how fast will the water exit the tank?

Ⓢ Electric Potential Energy

Three charges are lined up in a straight line at 1 mm intervals. Charge 1, in the center, has a charge of +1 μC. Charge 2, which is 1 mm to the right of Charge 1, has a charge of +2 μC. Charge 3, which is 1 mm to the left of Charge 1, has a charge of –3 μC. How much work was required to assemble this distribution of charges, assuming that the charges were initially separated by a distance of infinity? (Note: $k = 9 \times 10^9$ Nm²/C²)

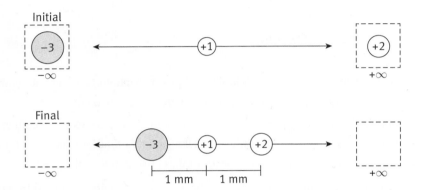

❶ At what distance would Charges 1 and 2 have an energy of interaction less than 1 mJ?

The phrase "Charges are initially separated by a distance of infinity" implies that the charges being discussed are separated by a great enough distance that the electrostatic potential energy between them is practically zero. However, what "practically zero" means depends on the needs of a given scientific experiment! For this question, an interaction on the scale of milli-Joules might be considered "practically zero." So, in order to approximate this notion of "infinite separation," we can use the electric potential energy equation to calculate the distance that would result in a very small potential energy, such as 1 mJ.

$$U = \frac{kQq}{r}$$

$$1 \cdot 10^{-3}\,\text{J} = \frac{k \times 1 \cdot 10^{-6}\,\text{C} \times 2 \cdot 10^{-6}\,\text{C}}{\text{distance (m)}}$$

$$\text{distance} = \frac{9 \cdot 10^9\,\text{N} \cdot \text{m}^2 \cdot \text{C}^{-2} \times 1 \cdot 10^{-6}\text{C} \times 2 \cdot 10^{-6}\text{C}}{1 \cdot 10^{-3}\,\text{J}}$$

$$\text{distance} = \frac{18 \cdot 10^{-3}\,\text{N} \cdot \text{m}^2}{1 \cdot 10^{-3}\,\text{J}} \rightarrow \text{distance} = 18\,\text{m}$$

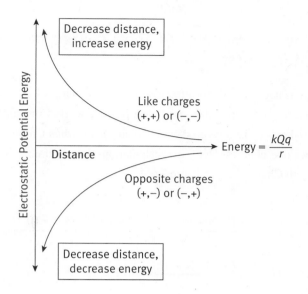

2 How much energy would be required to bring the second charge into place?

Moving the second charge into position requires positive work because like charges repel each other. Imagine holding two magnets, one in each hand, and bringing them close together, north pole to north pole—work is clearly required. In the same way that "like" poles of magnets repel, "like" charges repel as well. Whether with magnets or charges, overcoming repulsion requires positive work. Concretely, work is the difference in electrostatic potential energy when the charges are in their final position versus their initial position. Remember that at "infinity," the energy of interaction between two charges is approximately zero. So, to calculate the work, you can take the initial energy of the charges as "zero." This approximation ultimately reveals the reason why physicists are so fond of separating charges an "infinite" distance apart: when two charges start at an "infinite" distance apart (18 m or more in this problem), the work done bringing them together simply equals the energy of the charges in their final position. In order to bring Charge 1 and Charge 2 from infinity to 1 mm apart, the energy required would be:

$$U = \frac{9 \cdot 10^9 \,\text{N} \cdot \text{m}^2 \cdot \text{C}^{-2} \times 1 \cdot 10^{-6}\,\text{C} \times 2 \cdot 10^{-6}\,\text{C}}{1 \cdot 10^{-3}\,\text{m}}$$

$$U = \frac{18 \cdot 10^{-3}\,\text{N} \cdot \text{m}^2}{1 \cdot 10^{-3}\,\text{m}} = 18 \text{ J}$$

3 · Why would moving the third charge into position result in negative work?

The third charge is negative, so it is attracted to positive Charges 1 and 2. Recall that putting Charge 2 in proximity to Charge 1 took positive work. Conversely, the electrostatic potential energy is decreased, and thus negative, when Charge 3 is brought closer to the positive charges. Here, two calculations are required, since Charge 3 gets closer to both Charge 1 and Charge 2.

Charge 1

$$U = \frac{9 \cdot 10^9 \, N \cdot m^2 \cdot C^{-2} \times 1 \cdot 10^{-6} C \times -3 \cdot 10^{-6} C}{1 \cdot 10^{-3} m}$$

$$U = \frac{-27 \cdot 10^{-3} \, N \cdot m^2}{1 \cdot 10^{-3} m} = -27 \; J$$

Charge 2

$$U = \frac{9 \cdot 10^9 \, N \cdot m^2 \cdot C^{-2} \times 2 \cdot 10^{-6} C \times -3 \cdot 10^{-6} C}{2 \cdot 10^{-3} m}$$

$$U = \frac{-54 \cdot 10^{-3} \, N \cdot m^2}{2 \cdot 10^{-3} m} = -27 \; J$$

The work released by bringing Charge 3 into the system is –54 J.

Takeaways

To find the work required to assemble a distribution of charges, find the work required to place each charge individually, and then add them together. The work to place the first charge is always zero.

Things to Watch Out For

It is common to make errors with the sign convention on these types of problems. Keep this in mind when checking your work: like charges increase in potential energy as they are brought closer to each other; unlike charges decrease in potential energy as they are brought closer to each other.

4 · What is the net work done in arranging these charged particles?

Net work can be calculated by adding all the potential energies previously calculated. The energy of the interaction of the charges are: +18 joules (Charges 1 and 2), –27 joules (Charges 1 and 3), and –27 joules (Charges 2 and 3). The electrical potential energy of this system is therefore –36 joules. This conclusion reflects the fact that it would require positive work to separate the charges in this system. Or, in plain English: Despite the repulsion between the positive charges (Charges 1 and 2), the three charges are more stable together than they were when they were separated.

$$W_{system} = U_{1,2} + U_{1,3} + U_{2,3}$$

$$W_{system} = 18 \; J + (-27 \; J) + (-27 \; J) = -36 \; J$$

Related Questions

1. A +1 μC charge sits 1 cm from a −2 μC charge. How much work is done in tripling the distance between these charges?

2. How much work is done in assembling a square-shaped charge distribution with a side length of 1 μm if all of the charges have a charge of 5 nC?

3. Charges 1, 2, and 3 are lined up, in that order, at 1 mm intervals along the y-axis. Charge 1 has a charge of +4 μC. Charge 2 has a charge of −2 μC. Charge 3 has a charge of −3 μC. What is the change in potential energy of the system if Charge 1 is removed?

E Resistor Circuits

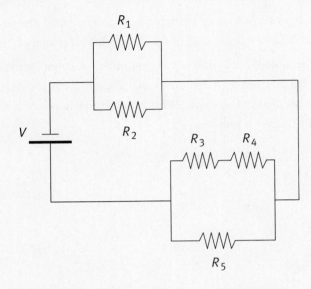

What is the current through R_1 in the circuit shown below?

(Note: $R_1 = 30\ \Omega$; $R_2 = 6\ \Omega$; $R_3 = 20\ \Omega$; $R_4 = 10\ \Omega$; $R_5 = 30\ \Omega$; $V = 30\ V$)

1 What is the equivalent resistance of the network?

The first part of this problem is to find the equivalent resistance of the entire circuit. This will take several steps. Begin by combining R_1 and R_2 using the equation for the equivalent resistance of two parallel resistors:

$$\frac{1}{R_{eq1}} = \frac{1}{R_1} + \frac{1}{R_2} = \frac{1}{30\ \Omega} + \frac{1}{6\ \Omega} = \frac{1}{30} + \frac{5}{30} = \frac{6}{30} = \frac{1}{5}$$

$$R_{eq1} = 5\ \Omega$$

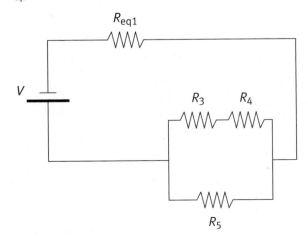

Now, combine R_3 and R_4, which are in series:

$$R_{eq2} = R_3 + R_4 = 20\,\Omega + 10\,\Omega$$

$$R_{eq2} = 30\,\Omega$$

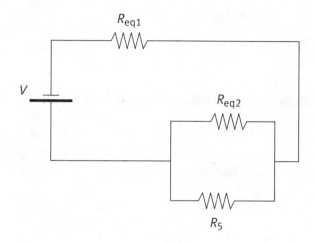

Now, combine the equivalent resistance R_{eq2} with R_5. These resistors are in parallel:

$$\frac{1}{R_{eq3}} = \frac{1}{R_{eq2}} + \frac{1}{R_5} = \frac{1}{30\,\Omega} + \frac{1}{30\,\Omega} = \frac{2}{30}$$

$$R_{eq3} = 15\,\Omega$$

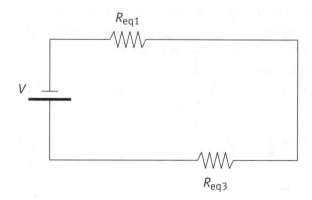

Finally combine R_{eq1} and R_{eq3}. These resistors are in series:

$$R_{eq4} = R_{eq1} + R_{eq3} = 5\,\Omega + 15\,\Omega$$

$$R_{eq4} = 20\,\Omega$$

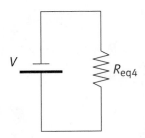

2 What is the current through the circuit?

The point of finding the equivalent resistance is so that we can find the current through the circuit. This is also often referred to as the current through or leaving the battery. Use Ohm's law to find the current from the source voltage and equivalent resistance:

$$V = IR \rightarrow I = \frac{V}{R} = \frac{30 \text{ V}}{20 \text{ }\Omega} = 1.5 \text{ A}$$

3 How could you expand the circuit and solve for individual components?

Now expand the circuit back out and apply what we know about resistors in series and parallel to find the current and voltage through individual resistors. All resistors in series must have the same current, so we know that the current from Step 2 must equal the current through R_{eq1} and R_{eq3}:

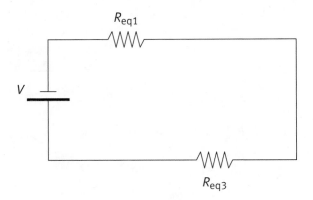

Use Ohm's law to find the voltage across R_{eq1}:

$$V_{eq1} = I_{eq1}R_{eq1}$$
$$= (1.5\,\text{A})(5\,\Omega)$$
$$= 7.5\,\text{V}$$

Because R_{eq1} is a parallel combination of R_1 and R_2, we know that R_1 and R_2 must have the same voltage as R_{eq1}. Any two (or more) circuit components in parallel must have the same voltage as the others:

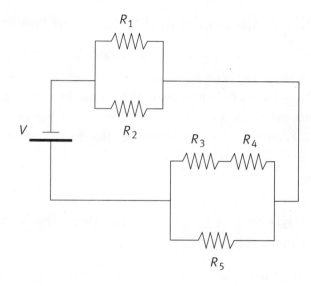

Takeaways

There is a standard process for solving these types of resistor problems:
1. Find the equivalent resistance by combining series and parallel resistors.
2. Find the current through the circuit.
3. Expand the circuit back out, step by step.
4. Apply knowledge of series and parallel resistors to find the quantity of interest.

Things to Watch Out For

The arithmetic for calculating the equivalent resistance in parallel can be confusing. Another version of the formula is $R_p = \dfrac{R_1 R_2}{R_1 + R_2}$. Note, however, that this only works for two resistors and cannot be easily expanded to account for three resistors.

Use the voltage and resistance of R_1 to find the current through R_1:

$$V_1 = I_1 R_1 \rightarrow I_1 = \frac{V_1}{R_1}$$
$$= \frac{7.5\,\text{V}}{30\,\Omega} = 0.25\,\text{A}$$

Related Questions

1. In the circuit shown in the original question, what is the voltage across R_4?

2. Four resistors are attached in parallel to a voltage supply of 9 V. If the resistances of the resistors are $10\,\Omega$, $20\,\Omega$, $30\,\Omega$, and $40\,\Omega$, what is the current through the battery?

3. There are six resistors in a circuit. R_1, R_2, R_3, R_4, and R_5 are all in parallel with each other, and they are all in series with R_6. If the current leaving the battery is 10 A and each of the resistors has a resistance of $1\,\Omega$ except for R_6, which is unknown, what is the resistance of R_6 if $V = 20\,\text{V}$?

E Doppler Effect

> Two cars, Car A and Car B, are moving toward each other, with each car traveling at $50\frac{m}{s}$ when Car B starts to beep its 475 Hz horn. Assuming that the speed of sound is $343\frac{m}{s}$, what is the wavelength of the horn as perceived by the driver of Car A?

1 What are the source and detector of the wave?

This problem has the classic setup of a Doppler effect question. Whenever you see a setup in which one object emits a wave and another object detects it, think about the Doppler effect—especially if those two objects are moving relative to each other. First, let's define our terms. The source is the object that emits the sound wave; this is Car B. The detector or observer is the object that detects the wave; this is the person in Car A.

2 What is the effect of detector velocity on the perceived frequency?

Every Doppler effect problem can be solved using the Doppler effect equation:

$$f' = f\frac{(v \pm v_D)}{(v \mp v_S)}$$

where f' is the observed frequency, f is the emitted frequency, v is the speed of sound in the medium, v_D is the speed of the detector, and v_S is the speed of the source. The most complicated aspects of using the Doppler effect equation are remembering where to place the variables and the sign convention. While we present the logic for the sign convention throughout this explanation, a mnemonic can be used: if memorized in this form, the top sign should be used when the detector or source is moving *toward* the other object, whereas the bottom sign should be used when the detector or source is moving *away* from the other object.

Let's start with the detector (the person in Car A). The detector is moving toward the source; therefore, the observed frequency, f', will be greater than if the detector were stationary. The numerator of the Doppler effect equation represents the motion of the detector; the fact that the detector is moving toward the source implies that a plus sign should be used in the numerator. Remember: the top sign in the numerator (plus) should be used when the detector is moving toward the source:

$$f' = f\frac{(v + v_D)}{(v \mp v_S)}$$

3 What is the effect of source velocity on the perceived frequency?

Now, let's consider the source (the horn on Car B). The source is moving toward the detector; therefore, the observed frequency, f', will be greater than if the source were stationary. The denominator of the Doppler effect equation represents the motion of the source; the fact that the source is moving toward the detector implies that a minus sign should be used in the denominator. Remember: the top sign in the denominator (minus) should be used when the source is moving toward the detector:

$$f' = f\frac{(v + v_D)}{(v - v_S)}$$

4 What is the perceived frequency?

We are given the following values:

$$f = 475\ \text{Hz}$$

$$v = \text{speed of sound} = 343\ \frac{\text{m}}{\text{s}}$$

$$v_D = v_S = 50\ \frac{\text{m}}{\text{s}}$$

Plugging these into the equation, we get:

$$f' = f\frac{(v + v_D)}{(v - v_S)}$$

$$= (475\ \text{Hz})\frac{\left(343\ \frac{\text{m}}{\text{s}} + 50\ \frac{\text{m}}{\text{s}}\right)}{\left(343\ \frac{\text{m}}{\text{s}} - 50\ \frac{\text{m}}{\text{s}}\right)}$$

$$= 475 \times \frac{393}{293} \approx 480 \times \frac{4}{3} = 640\ \text{Hz (actual} = 637.1\ \text{Hz)}$$

5 What is the wavelength of the perceived frequency?

This question asks for the wavelength of the observed wave, not its frequency. We can use the relationship between wave speed, frequency, and wavelength to determine the answer:

$$v = f\lambda \rightarrow \lambda = \frac{v}{f}$$

$$\lambda = \frac{343\ \frac{\text{m}}{\text{s}}}{637\ \text{Hz}} \approx 0.5\ \text{m}\left(\text{actual} = 0.54\ \text{m}\right)$$

Takeaways

As the detector approaches the source, the observed frequency will be higher than the emitted frequency, and when the detector moves away from the source, the observed frequency will be lower than the emitted frequency. The same rule applies to the motion of the source. Therefore, when determining the right sign convention of the Doppler equation, use the sign in the equation that will yield the appropriate observed frequency.

Things to Watch Out For

Always separate Doppler effect questions into two parts: the effect of the source and the effect of the observer. Be careful in problems where the object and the source are moving in the same direction. If the source is ahead of the detector, the source will be moving away from the detector while the detector is moving toward the source, regardless of their relative speeds. If the detector is ahead of the source, the detector will be moving away from the source while the source is moving toward the detector.

Related Questions

1. Suppose a policeman running at $5\,\frac{m}{s}$ is firing his gun at a rate of 20 bullets per minute while chasing a bank robber who is driving away at $50\,\frac{m}{s}$. At what rate do the bullets reach the bank robber? (Note: Use $500\,\frac{m}{s}$ for the speed of a bullet.)

2. A bungee jumper yells at 350 Hz as he falls off a bridge toward a river at a rate of $20\,\frac{m}{s}$. What are the frequencies heard by the observers on the bridge and a boat on the river? (Note: The speed of sound is $343\,\frac{m}{s}$.)

3. Some animals use echolocation to navigate three-dimensional space and find prey. During echolocation, what is the detector and what is the source? What signs would be used in the Doppler effect equation in this case?

High-Yield Problem-Solving Guide questions continue on the next page. ▶ ▶ ▶

Ⓢ Geometrical Optics

Cataracts are responsible for roughly half of blindness cases worldwide. Though this condition is marked by a clouding of the lens, cataracts may occasionally also lead to variation in the refractive index of the lens. This condition creates additional problems similar to those seen when the axial length of the eye is altered over time.

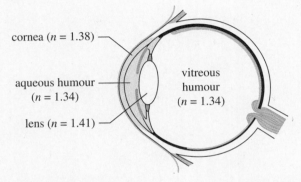

cornea ($n = 1.38$)

aqueous humour ($n = 1.34$)

lens ($n = 1.41$)

vitreous humour ($n = 1.34$)

What condition might arise from the lens having a greater density than normal, and how might this condition be corrected?

① How do variations to the axial length cause myopia and hyperopia?

Once the light crosses the lens–vitreous humour interface, it continues its path in a straight line until it reaches the retina. When the only optical abnormality is alteration of the axial length, the converging system at the front of the eye is unaltered. Consequently, in an eye that is too short, the rays will converge behind the retina, while in an abnormally long eye, the image will form in front of the retina.

When rays converge behind the retina, the image is formed beyond the retina and cannot be properly seen. Converging (convex) lenses are corrective in hyperopia, as they force the light to converge onto the retina, rather than behind it. Similarly, when the rays converge in front of the retina, they are converging too quickly, making diverging (concave) lenses corrective in myopia.

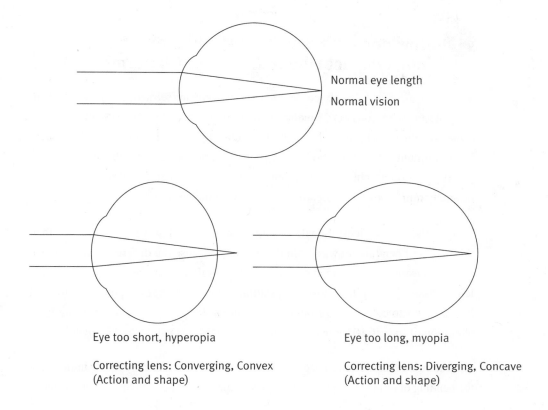

Normal eye length
Normal vision

Eye too short, hyperopia

Correcting lens: Converging, Convex
(Action and shape)

Eye too long, myopia

Correcting lens: Diverging, Concave
(Action and shape)

② What is the general relationship between an angle and its sine value?

This relationship is particularly important for Snell's law. Remembering that $\sin 0° = 0$ and $\sin 90° = 1$, it is clear that from 0 to 90 degrees, as the angle value increases, so does its sine function value.

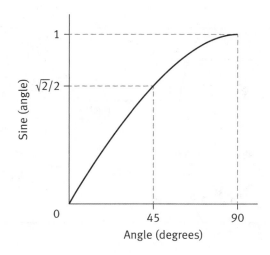

③ How does the lens having a higher refractive index impact where the image will form?

The lens's increased density confers it a slightly greater refractive index. There are two interfaces to consider: aqueous humour–lens and lens–vitreous humour. Because the abnormality is at the level of the lens, light will be refracted in the same fashion up until it reaches the lens. A greater refractive index of the lens indicates that it will refract light more. The first equation tells us that the refracted angle will be smaller, indicating light bending more toward the normal.

When light exits the lens, the vitreous humour, having a lower refractive index than the lens, will bend light away from the normal. Because the difference in refractive indices is even wider in the abnormal eye, light will be bent further away from the lens–vitreous humour than normal, resulting in more converging light rays. This will lead the rays to converge in front of the retina. In fact, the condition resulting from alteration of the refractive indices of the ocular media is called "index myopia."

Keep in mind that the difference in refractive index is not the only factor influencing the bending of light! The lens acts as a typical converging lens, having a convex shape.

$$n_{aq.h.} \times \sin \theta_{aq.h.} = n_{lens} \times \sin \theta_{lens}$$
$$n_{lens} \times \sin \theta_{lens} = n_{vit.h.} \times \sin \theta_{vit.h.}$$

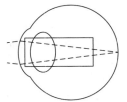

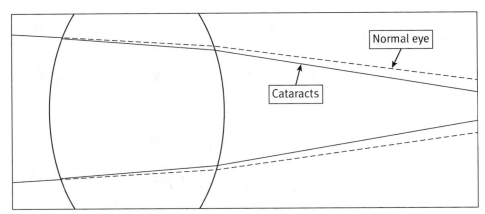

Normal eye

Cataracts

Related Questions

1. A firefighter shines her flashlight through a smoky room at a 45° angle to a window. What angle does the beam of light make with the pane of glass on the outside? (Note: The indices of refraction are as follows: air, $n = 1$; glass, $n = 1.5$; smoke, $n = 1.1$.)

2. A jeweler is appraising the stone on a ring. He aims a beam of light 35° from the normal at a flat edge of the gem. What angle would he observe in the gem if the stone were diamond ($n = 2.4$)? What angle would he observe if the stone were zircon ($n = 1.9$)?

3. If light were traveling from salt water into air, what would be the associated critical angle?

Takeaways

Complex problems can be solved with a little bit of ingenuity. Snell's law is a simple concept and is likely to be one of the first steps in any problem dealing with the refraction of light. When light passes through multiple layers, the final angle can be determined merely by comparing the first and final media.

Things to Watch Out For

Light reflects off of media boundaries, too. Because of this, the refracted ray of light is less intense than the incident ray. Different wavelengths also have slightly different indices of refraction—this difference is what causes light dispersion.

E Nuclear Reactions

A neutral U-238 atom absorbs a neutron and then immediately undergoes two alpha decays, three beta decays, one positron decay, and two gamma decays (not necessarily in that order). Describe the resulting nucleus using isotopic notation. The resulting atom is an isotope of what element?

1 What are the mass and atomic numbers of the parent nucleus?

The notation U-238 indicates a uranium atom with a mass number (A) of 238. The letter is always the element's symbol, as listed on the Periodic Table. From the Periodic Table, the atomic number of uranium (Z) is 92: all uranium atoms, by definition, have 92 protons.

The mass number, A, is the sum of the number of protons and neutrons. The mass number is given in the question because a given element may have many different isotopes, which all have the same number of protons but vary in the number of neutrons they contain.

Isotopes may be written in shorthand notation as $^A_Z X$, where X is the element's symbol, A is the mass number, and Z is the atomic number. Here, this uranium isotope could be written as $^{238}_{92} U$.

2 What is the result of neutron absorption?

A neutron has a mass number of 1 and an atomic number of 0 because it has no protons. In these problems, make sure to balance both the atomic numbers and the mass numbers in the decay equation:

$$^A_Z X + {}^1_0 n \rightarrow {}^{A+1}_Z X$$
$$^{238}_{92} U + {}^1_0 n \rightarrow {}^{239}_{92} U$$

The result of the absorption is the formation of a different isotope of uranium.

3 What is the result of the alpha decays?

An alpha particle is a helium nucleus: two protons and two neutrons. Thus, for a single alpha decay, A decreases by 4 and Z decreases by 2. Because there are two alpha decays, multiply these numbers by two to find the change in the values of Z and A:

$$_{Z}^{A}X \rightarrow {}_{Z-2}^{A-4}X + {}_{2}^{4}\alpha$$
$$_{92}^{239}U \rightarrow {}_{88}^{231}Ra + 2\,{}_{2}^{4}\alpha$$

4 What is the result of positron decay?

Beta decay comes in two forms: beta-minus and beta-plus (also called positron emission). In β^{-} decay, a neutron decays into a proton, a β-particle, and an antineutrino. The β-particle, which is nothing more than an electron, is ejected from the nucleus. Thus, the mass number stays the same, while the atomic number of the daughter nucleus increases by 1. There are three beta decays in this question:

$$_{Z}^{A}X \rightarrow {}_{Z+1}^{A}X + {}_{-1}^{0}\beta$$
$$_{88}^{231}Ra \rightarrow {}_{91}^{231}Pa + 3\,{}_{-1}^{0}\beta$$

5 What is the result of the gamma decays?

Positron decay is the exact opposite of beta decay and is often called β^{+} decay. In positron decay, a proton decays into a neutron, a positron, and a neutrino. The positron (an anti-electron—a particle with a positive charge and negligible mass) is ejected from the nucleus. Thus, the mass number stays the same, and the atomic number decreases by 1:

$$_{Z}^{A}X \rightarrow {}_{Z-1}^{A}X + {}_{+1}^{0}\beta$$
$$_{91}^{231}Pa \rightarrow {}_{90}^{231}Th + {}_{+1}^{0}\beta$$

Takeaways

It is important to know the decay types and the identities of the decay particles themselves. Once you have these down, these problems are relatively simple.

6 What is the result of the gamma decays?

Things to Watch Out For

Fission occurs when a large nucleus splits into smaller nuclei. Fusion is the combination of smaller nuclei to form a larger particle. Transmutation, or radioactive decay resulting in a change of atomic number, is a specific type of fission reaction. β-particles (electrons) and positrons are very easy to confuse; make sure that you understand the differences between them.

In gamma decay, a gamma ray (an electromagnetic wave) is ejected from the nucleus. There is no change in atomic number or mass number:

$$_{Z}^{A}X^{*} \rightarrow {}_{Z}^{A}X + \gamma$$
$$_{90}^{231}Th^{*} \rightarrow {}_{90}^{231}Th + 2\gamma$$

The daughter nucleus of these eight decay processes is a thorium-231 atom, with 90 protons and $231 - 90 = 141$ neutrons.

Related Questions

1. Is it possible for neptunium to decay into an isotope of lead through a series of alpha decays?

2. How many β-particles are ejected when polonium-214 decays into radon-214?

3. Uranium-226 decays radioactively into radon-218 through two positron decays and an unknown number of alpha decays. How many alpha particles must be emitted in this reaction?

High-Yield Problem-Solving Guide questions continue on the next page. ▶ ▶ ▶

Key Concepts

E Dimensional Analysis

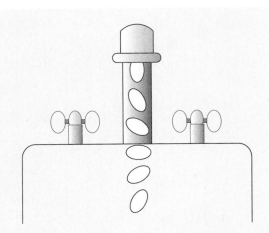

Water is dripping from a leaky faucet. As the drops fall, they oscillate as shown in the diagram above. Given that the frequency depends only on the surface tension T of the water (measured in $\frac{N}{m}$), the radius of the drops r, and the density of the water, use dimensional analysis to find a proportionality expression for the frequency f of the drops.

1 What are the relevant physical quantities, and what are their SI units?

As given in the problem, surface tension has units of newtons per meter. We can convert this to SI units by remembering that $1\ N = 1\frac{kg \cdot m}{s^2}$. Density can be given in the units of $\frac{kg}{m^3}$. Radius is a distance with the units of meters. Frequency has the units of Hz, or $\frac{1}{s}$. In summary:

$$T : \frac{N}{m} = \frac{kg}{s^2}$$

$$\rho : \frac{kg}{m^3}$$

$$r : m$$

$$f : Hz = \frac{1}{s}$$

2 What is the hypothetical frequency formula?

The frequency is related to the surface tension, T, the radius, r, and the density, ρ. Write an equation for these using variables (like x, y, and z) as exponents:

$$f = kT^x r^y \rho^z$$

where k is a unitless constant

3 How do the units plug into the hypothetical formula?

Plug the units into the hypothetical formula from Step 2:

$$f = kT^x r^y \rho^z$$

$$\frac{1}{s} = \left[\frac{kg}{s^2}\right]^x [m]^y \left[\frac{kg}{m^3}\right]^z$$

4 What are the values of the hypothetical exponents x, y, and z?

We know that the units on the left must equal the units on the right. Only the first term on the right side contains seconds, so we know x must be $\frac{1}{2}$; thus, we end up with $\frac{1}{s}$ on both sides.

Only the first and third terms contain kilograms; there are no units of kilograms on the left side of the equation. Therefore, the last term (density) must have an exponent of $z = -\frac{1}{2}$ so that the units of kilograms divide on the right side of the equation.

Finally, only the second and third terms contain meters; there are no units of meters on the left side of the equation. If the value of $z = -\frac{1}{2}$, then the middle term (radius) must have an exponent of $y = -\frac{3}{2}$ so that the units of meters divide on the right side of the equation.

5 What simple equation can you generate from these exponents?

All of our exponents are fractions, which means that a radical will be involved in the equation. Remember that a quantity raised to a negative exponent is the same as dividing by the same quantity raised to a positive exponent of the same magnitude. Putting this together, we can determine the equation:

$$f = kT^{\frac{1}{2}}r^{-\frac{3}{2}}\rho^{-\frac{1}{2}}$$

$$= k\frac{T^{\frac{1}{2}}}{r^{\frac{3}{2}}\rho^{\frac{1}{2}}}$$

$$= k\sqrt{\frac{T}{r^{3}\rho}}$$

Related Questions

1. What are the SI units of G in the universal law of gravitation?

2. An electric dipole is initially at rest in a uniform electric field. The torque provided by the electric field causes the dipole to oscillate back and forth. For the period of motion, the physicist derives the formula $T = kE^{\frac{1}{2}}$, where T is the period, E is the electric field, and k is a quantity with the appropriate units. Is this equation physically reasonable?

3. What are the units for the permeability of free space, μ_0?

High-Yield Problem-Solving Guide questions continue on the next page. ▶ ▶ ▶

E Study Design

A student wishes to document the effects of two different drugs in hypertensive patients. He obtains permission to contact all of the patients on propranolol and all of the patients on diltiazem from a certain pharmacy, and sends out written surveys asking for the patient's most recent blood pressure reading, age, and gender. Forty percent of respondents return their surveys. The student then compiles the diastolic blood pressure readings from each group. The mean systolic blood pressure in the propranolol group is 129 mmHg, and the mean in the diltiazem group is 127. The student concludes that diltiazem should be used as a first line treatment, rather than propranolol. Identify the errors that this student has made or the additional information that is needed to support his conclusion.

1 Where might error be introduced in the research design?

Break the components of any study down into phases; the list of question design, experimental method, data collection, data analysis, conclusions, and execution is generally a good place to start. Next, determine where error is likely to be introduced. Question design is not likely to lead to errors, but should be kept in mind during a discussion of the experimental method. Data collection is also relatively unlikely to lead to erroneous conclusions (although it can with systematic bias and inaccurate instruments) and execution is almost always free from errors that impact the study. Therefore, we should focus our evaluations of most studies on the experimental method, data analysis, and conclusions.

2 How does known execution of each step compare to ideal study design protocol?

To determine what errors have been made, we should compare the study to an ideal study of the same topic. Keep in mind that not all deviations from the ideal study are necessarily errors; some deviations may be deemed necessary because of practicality or financial concerns that may not substantially affect the data.

When considering the experimental method, one should always compare the study to the gold standard in biomedical research: the double-blind randomized controlled trial. The study that was conducted is not randomized, and does not have pre- and post-intervention data. The study was conducted from a single location and used self-report as the measurement instrument.

For data analysis, consider potential sources of bias. Studies with high response rates and sample sizes are ideal because they minimize selection bias. In this study, there was a very low response rate, which may have contributed to selection bias—only those who responded to the medication (or did not respond to the medication) may have replied to the survey, for example. Only one of the four collected variables was analyzed, and statistical findings were not provided. Whenever multiple variables may impact the measured outcomes, they should be considered as part of the analysis, and determinations of significance as well as p-values should be reported. This student did not account for other potential variables, such as severity of hypertension; demographic differences such as race, socioeconomic status, or access to healthcare; allergies to other study medications; and comorbid medical conditions.

To act upon a study, it should meet the requirements of the FINER approach and Hill's criteria. It should also demonstrate both clinical and statistical significance. In this study, we do not have sufficient information to draw the conclusions that the author makes for the reasons given above.

3 Is the flaw one of omission or design, and how can it be corrected?

The study is designed poorly, and requires a complete revision. However, there are several data points that could improve its relevance. An analysis that includes relevant p-values and adjusts for the collected variables could improve this study. Additionally, information about the clinical value of the different blood pressures would be worthwhile for study analysis in supporting the author's conclusion.

Related Questions

1. It is found that the number of shark attacks appears to correlate with the national gross sales of ice cream on a monthly basis. Why would it be a flawed conclusion to state that there is a causal relationship between purchasing ice cream and a shark attack?

2. How is respect for persons maintained in a study about the effects of electroconvulsive therapy?

3. How is causality established in an experimental study *vs.* an observational study?

Takeaways

Research errors are pervasive and compound one another. It is not enough to look for sample size or an overt bias because there may be issues that arise before any data is even collected.

Things to Watch Out For

Statistical significance and clinical significance are often used interchangeably in common parlance, but they are two different topics. In study analysis we must look for errors in interpretation of data and experimental design, not just data analysis.

Key Concepts

Physics and Math Chapter 12
Measures of central tendency
Measures of distribution
Charts, tables, and graphs
Comparative statistics

E Graphical Analysis

Given the following data sets for age of mortality in these imaginary countries, construct a series of box-and-whisker plots. Use these plots to determine which countries have the best and worst life expectancies:

Physia	Orgostan	United Biology	GCS
69	65	91	56
74	87	56	89
52	55	82	65
63	69	44	47
70	66	62	85
71	59	53	39
65	72	68	98
81	72	85	71
64	58	65	52
70	71	59	65

1 How would you order the data set? What are the mean, median, and interquartile ranges?

Median, mode, range, quartiles, and interquartile range (IQR) are easiest to calculate if the data are ordered, so this is the first step in generating comparisons. The reordered data are shown below:

Physia	Orgostan	United Biology	GCS
52	55	44	39
63	58	53	47
64	59	56	52
65	65	59	56
69	66	62	65
70	69	65	65
70	71	68	71
71	72	82	85
74	72	85	89
81	87	91	98

Now, it is possible to calculate the median, mean, and quartiles for this data. The median will be the average of the fifth and sixth data points for each set. Thus the medians are 69.5, 67.5, 63.5, and 65 for Physia, Orgostan, United Biology, and GCS, respectively.

We will calculate the mean by summing each column and dividing by ten. The means are 67.9, 67.4, 66.5, and 66.7 for Physia, Orgostan, United Biology, and GCS, respectively.

To find the quartile positions, we multiply the number of data points by $\frac{1}{4}$ and by $\frac{3}{4}$. If these are whole numbers, the quartiles are the averages of these data values and the ones above them. If they are not, then round up to find the positions of the quartiles. Here, $10 \times \frac{1}{4} = 2.5$ (round up to 3), and $10 \times \frac{3}{4} = 7.5$ (round up to 8). These data points are highlighted in green the following chart, with the median, mean, and IQR listed below the data.

Physia	Orgostan	United Biology	GCS
52	55	44	39
63	58	53	47
64	59	56	52
65	65	59	56
69	66	62	65
70	69	65	65
70	71	68	71
71	72	82	85
74	72	85	89
81	87	91	98

	Physia	Orgostan	United Biology	GCS
Median	69.5	67.5	63.5	65
Mean	67.9	67.4	66.5	66.7
IQR	7	13	26	33

Already, this data is more accessible than the initial presentation, but using a graphical form is even easier to interpret.

2 How could this data be represented in a box-and-whisker plot?

Using an axis that contains the minimum and maximum data points (39 and 98, respectively), construct a box-and-whisker plot that is divided according to the quartiles and median for each data set. The box plots can be constructed with or without outliers. Including outliers is useful information, but does require additional mathematics:

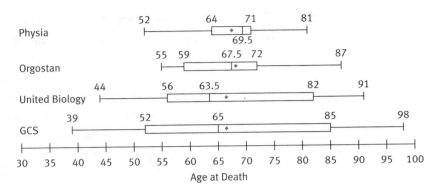

An outlier is any data point that is more than 1.5 × IQR away from the closest quartile. The one outlier in this data set is highlighted in red in the data table.

3 What comparisons can you make using the box-and-whisker plot?

Using the box-and-whisker plots, determine how countries compare to one another. For example, three quarters of the people in Physia live longer than half of the people in United Biology, so they have a longer life expectancy despite having similar mean and median values. In this data set, Physia has the best longevity outcomes when compared to the other groups; both United Biology and GCS are very similar and have the poorest outcomes.

Related Questions

1. Provide at least three criticisms of the following chart (don't be afraid to state the obvious!):

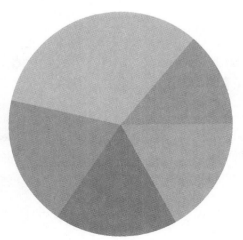

2. What are the appropriate first steps when analyzing any graph or table?

3. Determine the equations for the two lines below. At what point do they intersect?

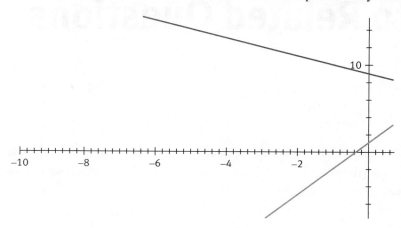

Solutions to Related Questions

1. Inclined Plane

1. The only two forces acting on this block in the parallel dimension are the force of static friction pointing up the plane and the parallel component of gravity pointing down the plane. If the block is at rest, then it is in translational equilibrium and the magnitudes of these two forces must be equal:

$$f_s = F_{g,\parallel}$$
$$\mu_s N = mg\sin\theta$$
$$\mu_s mg\cos\theta = mg\sin\theta$$
$$\mu_s = \frac{\sin\theta}{\cos\theta} = \tan\theta = \tan 45° = 1$$

2. This question can be solved with kinematics equations, but it is actually easier to use conservation of energy to arrive at the answer. If the ramp is frictionless, then there are no nonconservative forces and all of the kinetic energy will be converted to gravitational potential energy. The initial gravitational potential energy can be assigned the value of zero because the question asks for the height above the initial point. The final kinetic energy is also zero because the block has zero speed when it reaches its highest point. Therefore, the initial kinetic energy and final gravitational potential energy are equal:

$$K_1 = U_2$$
$$\frac{1}{2}mv_1^2 = mgh_2$$
$$h_2 = \frac{v_1^2}{2g} \approx \frac{\left(2\,\frac{m}{s}\right)^2}{2\times 10\,\frac{m}{s^2}} = 0.2\text{ m}$$

3. Start by determining the acceleration. Because the only force in the parallel direction is the parallel component of gravity, the acceleration will be $g\sin\theta = 10\,\frac{m}{s^2} \times 0.5 = 5\,\frac{m}{s^2}$. Now, use a kinematics equation:

$$v = v_0 + at$$
$$= 0 + \left(5\,\frac{m}{s^2}\right)(5s) = 25\,\frac{m}{s}\left(\text{actual} = 24.5\,\frac{m}{s}\right)$$

2. Projectile Motion and Air Resistance

1. The rock starts with zero kinetic energy and a gravitational potential energy of mgh_i. It ends with a kinetic energy of $\frac{1}{2}mv_f^2$ and zero gravitational potential energy. We are asked how much energy is lost, which would be the difference between these two terms:

$$W_{\text{nonconservative}} = E_i - E_f$$
$$= mgh_i - \frac{1}{2}mv_f^2$$
$$= m \times 10\,\frac{m}{s^2} \times 100\,m - \frac{1}{2} \times m \times \left(30\,\frac{m}{s}\right)^2$$
$$= 1000\,m - 450\,m = 550\,m$$

We cannot say an exact value for the amount of energy lost without knowing the mass of the rock, but $550m$ joules of energy have been lost, where m is the mass of the rock. We could also say that $\frac{550\,m}{1000\,m} = 55\%$ of the energy has been lost.

2. Let's call the speed of the slower rock v. This means that the speed of the faster rock is $1.3\,v$. The difference in kinetic energy between the rocks is:

$$\frac{1}{2}m(1.3v)^2 - \frac{1}{2}mv^2 = 0.69 \times \frac{1}{2}mv^2$$

The question asks how much more kinetic energy the faster rock has. We cannot give an exact value, but can compare it to the kinetic energy of the slower rock—the faster rock has 69% more kinetic energy than the slower rock.

3. First, find the difference in energy between the initial and final point:

$$W_{\text{nonconservative}} = E_i - E_f$$
$$= \frac{1}{2}mv_i^2 - mgh_f$$

Now, divide that by the initial energy:

$$\frac{\frac{1}{2}mv_i^2 - mgh_f}{\frac{1}{2}mv_i^2} = \frac{v_i^2 - 2gh_f}{v_i^2}$$

$$\approx \frac{\left(30\,\frac{m}{s}\right)^2 - 2 \times 10\,\frac{m}{s^2} \times 44.1\,m}{\left(30\,\frac{m}{s}\right)^2} = \frac{900 - 882}{900} = \frac{18}{900} = 0.02 = 2\%\ (\text{actual} = 4\%)$$

3. Thermodynamics

1. A minimum of three processes is required to have a closed cycle. This requires either an adiabatic or an isothermal process as well as an isobaric process (one with a constant pressure) and an isovolumetric process (one with a constant volume; also called an isochoric process). This creates a shape like a right triangle on a pressure–volume graph. While three-process cycles are seen, it is most common for closed-cycle processes to have four steps as seen in the Carnot cycle.

2. Entropy change can be given by the equation $\Delta S = \frac{Q_{\text{rev}}}{T}$. During an adiabatic process, $Q = 0$, thus $\Delta S = 0$. During an adiabatic process, there is no entropy change.

3. The liquid (usually mercury in older thermometers) is undergoing expansion. Because there is a change in temperature, this process is not isothermal. Because the change in temperature is accompanied by heat exchange, it is not adiabatic

either. This process is likely to be isobaric, taking place under constant pressure. The water is cooling very slightly as a result of the distribution of heat to the thermometer, and consequently will experience some compression.

4. Fluid Dynamics

1. If the pipe is horizontal, then $h_1 = h_2$ in Bernoulli's equation. This simplifies the equation:

$$P_1 + \frac{1}{2}\rho v_1^2 = P_2 + \frac{1}{2}\rho v_2^2$$

We are told that $P_1 = 3P_2$. We can now solve for the linear flow speed at Point 1 in terms of the linear flow speed at Point 2:

$$3P_2 + \frac{1}{2}\rho v_1^2 = P_2 + \frac{1}{2}\rho v_2^2$$
$$\frac{1}{2}\rho v_1^2 = \frac{1}{2}\rho v_2^2 - 2P_2$$
$$v_1^2 = v_2^2 - \frac{4P_2}{\rho}$$
$$v_1 = \sqrt{v_2^2 - \frac{4P_2}{\rho}}$$

2. This question can be solved using the continuity equation. The radius at Point A is twice the radius at Point B, meaning that cross-sectional area at Point A is four times that at Point B:

$$A_A v_A = A_B v_B$$
$$\left(\pi r_A^2\right)v_A = \left(\pi r_B^2\right)v_B$$
$$\frac{v_A}{v_B} = \left(\frac{r_B}{r_A}\right)^2 = \left(\frac{1}{2}\right)^2 = \frac{1}{4}$$

Thus, $v_A : v_B = 1:4$.

3. We can solve this question with Bernoulli's equation again. In this case, the pressure is the same (atmospheric pressure) because the tank is open to the air. We can again assume the linear flow speed at the surface is essentially zero. Thus, Bernoulli's equation simplifies to:

$$\rho g h_1 = \frac{1}{2}\rho v_2^2 + \rho g h_2$$
$$\frac{1}{2}v_2^2 = g h_1 - g h_2$$
$$v_2 = \sqrt{2g(h_1 - h_2)} = \sqrt{2 \times 10\,\frac{m}{s^2} \times (1\,m - 0.5\,m)} = \sqrt{10} \approx 3.16\,\frac{m}{s}$$

5. Electric Potential Energy

1. The amount of work done is equal to the change in electrical potential energy:

$$\Delta U = \frac{kQq}{r_2} - \frac{kQq}{r_1} = kQq\left(\frac{1}{r_2} - \frac{1}{r_1}\right)$$

$$= \left(9 \times 10^9 \, \frac{\text{N} \cdot \text{m}^2}{\text{C}^2}\right)\left(1 \times 10^{-6}\text{C}\right)\left(-2 \times 10^{-6}\text{C}\right)\left(\frac{1}{3 \times 10^{-2}\text{m}} - \frac{1}{1 \times 10^{-2}\text{m}}\right)$$

$$= -18 \times 10^{-3}\left(\frac{1}{3 \times 10^{-2}} - \frac{3}{3 \times 10^{-2}}\right)$$

$$= -18 \times 10^{-3}\left(-\frac{2}{3 \times 10^{-2}}\right) = 1.2 \text{ J}$$

2. The amount of work can be determined by summing the work to move each charge into the configuration:

$$W_1 = 0 \, V$$

$$W_2 = \frac{kQq}{r_1} = \frac{kQq}{1 \times 10^{-6}\text{m}}$$

$$W_3 = \frac{kQq}{r_1} + \frac{kQq}{r_2} = \frac{kQq}{1 \times 10^{-6}\text{m}} + \frac{kQq}{1.4 \times 10^{-6}\text{m}}$$

$$W_4 = \frac{kQq}{r_1} + \frac{kQq}{r_2} + \frac{kQq}{r_3} = \frac{2kQq}{1 \times 10^{-6}\text{m}} + \frac{kQq}{1.4 \times 10^{-6}\text{m}}$$

$$W_{total} = \frac{4kQq}{1 \times 10^{-6}} + \frac{2kQq}{1.4 \times 10^{-6}} = \frac{kQq}{10^{-6}}\left(4 + \frac{2}{1.4}\right)$$

$$\approx \frac{\left(9 \times 10^9 \, \frac{\text{N} \cdot \text{m}^2}{\text{C}^2}\right)\left(5 \times 10^{-9}\text{C}\right)\left(5 \times 10^{-9}\text{C}\right)}{10^{-6}\text{m}}(5.4)$$

$$\approx 225 \times 10^{-3} \times 5.4 \approx 1.22 \text{ J}$$

3. The initial potential energy of the system is:

$$U_i = \frac{kq_1q_2}{r_{1,2}} + \frac{kq_1q_3}{r_{1,3}} + \frac{kq_2q_3}{r_{2,3}}$$

$$= \left(9 \times 10^9 \, \frac{\text{N} \cdot \text{m}^2}{\text{C}^2}\right)\left[\frac{\left(4 \times 10^{-6}\text{C}\right)\left(-2 \times 10^{-6}\text{C}\right)}{1 \times 10^{-3}\text{m}} + \frac{\left(4 \times 10^{-6}\text{C}\right)\left(-3 \times 10^{-6}\text{C}\right)}{2 \times 10^{-3}\text{m}} + \frac{\left(-2 \times 10^{-6}\text{C}\right)\left(-3 \times 10^{-6}\text{C}\right)}{1 \times 10^{-3}\text{m}}\right]$$

$$= \left(9 \times 10^9\right)\left[-8 \times 10^{-9} - 6 \times 10^{-9} + 6 \times 10^{-9}\right]$$

$$= -72 \text{ J}$$

The final potential energy of the system is:

$$U_f = \frac{kq_2q_3}{r_{2,3}}$$

$$= \left(9 \times 10^9 \, \frac{\text{N} \cdot \text{m}^2}{\text{C}^2}\right)\left[\frac{\left(-2 \times 10^{-6}\text{C}\right)\left(-3 \times 10^{-6}\text{C}\right)}{1 \times 10^{-3}\text{m}}\right]$$

$$= \left(9 \times 10^9\right)\left(6 \times 10^{-9}\right)$$

$$= 54 \text{ J}$$

The change in electrical potential energy is therefore $54 - (-72) = +126$ J.

6. Resistor Circuits

1. We know that the current through the circuit is 1.5 A and the equivalent resistance of the portion of the circuit including R_3, R_4, and R_5 is 15 Ω. Thus, the voltage across this part of the circuit is $V = IR = (1.5 \text{ A})(15 \text{ Ω}) = 22.5$ V. The same voltage drop will be seen across the branch including R_3 and R_4. The total voltage drop will be the sum of the voltage drop across R_3 plus the voltage drop across R_4. The current is the same through each resistor, so the ratio of their resistances will be the same as the ratio of their voltage drops. There is a 2:1 ratio of resistance, so the voltage drop must be 15 V across R_3 and 7.5 V across R_4.

2. Use the equation for equivalent resistance of resistors in parallel:

$$\frac{1}{R_p} = \frac{1}{10\,\Omega} + \frac{1}{20\,\Omega} + \frac{1}{30\,\Omega} + \frac{1}{40\,\Omega}$$

$$= \frac{12 + 6 + 4 + 3}{120} = \frac{25}{120} = \frac{5}{24}$$

$$R_p = \frac{24}{5} = 4.8\,\Omega$$

Now, use Ohm's law:

$$V = IR \rightarrow I = \frac{V}{R}$$

$$I = \frac{9 \text{ V}}{4.8\,\Omega} \approx 2 \text{ A (actual} = 1.88 \text{ A)}$$

3. We know the current and voltage of the circuit, which allows us to find the equivalent resistance:

$$V = IR \rightarrow R = \frac{V}{I} = \frac{20 \text{ V}}{10 \text{ A}} = 2\,\Omega$$

Now determine the equivalent resistance of R_1 through R_5:

$$\frac{1}{R_p} = \frac{1}{R_1} + \frac{1}{R_2} + \frac{1}{R_3} + \frac{1}{R_4} + \frac{1}{R_5} = \frac{5}{1\,\Omega}$$

$$R_p = 0.2\,\Omega$$

Because R_1 through R_5 are in series with R_6, the resistance of R_6 must be $2 - 0.2 = 1.8$ Ω.

7. Doppler Effect

1. We are given all of the necessary variables to plug into the Doppler effect equation. Here, f = 20 bullets per minute, $v = 500\frac{m}{s}$, $v_D = 50\frac{m}{s}$, $v_S = 5\frac{m}{s}$. The detector (the bank robber) is moving away from the source, so a minus sign is used in the numerator. The source (the policeman's gun) is moving toward the detector, so a minus sign is used in the denominator, as well:

$$f' = f\frac{(v - v_D)}{(v - v_S)}$$

$$= \left(20\,\frac{bullets}{min}\right)\frac{\left(500\,\frac{m}{s} - 50\,\frac{m}{s}\right)}{\left(500\,\frac{m}{s} - 5\,\frac{m}{s}\right)} = 20 \times \frac{450}{495} \approx 20 \times \frac{9}{10} = 18\,\frac{bullets}{min}$$

2. In this question, f = 350 Hz, $v = 343\,\frac{m}{s}$, $v_D = 0\,\frac{m}{s}$, and $v_S = 20\,\frac{m}{s}$. There are two scenarios. In one case, the source is moving away from observers on the bridge, so a plus sign would be used in the denominator for this Doppler effect equation. In the other case, the source is moving toward observers on the river, so a minus sign would be used in the denominator for this Doppler effect equation:

$$f'_{bridge} = f\frac{v}{(v + v_S)} = (350\,Hz)\frac{343\,\frac{m}{s}}{\left(343\,\frac{m}{s} + 20\,\frac{m}{s}\right)} = 350 \times \frac{343}{363} = 331\,Hz$$

$$f'_{river} = f\frac{v}{(v - v_S)} = (350\,Hz)\frac{343\,\frac{m}{s}}{\left(343\,\frac{m}{s} - 20\,\frac{m}{s}\right)} = 350 \times \frac{343}{323} = 372\,Hz$$

3. During echolocation, the same animal serves as both the source and the detector. The sound wave bounces off of some surface and returns to the same animal that emitted the original frequency. In this case, as the animal flies toward some object, it could be said that the source is moving toward the detector and the detector is also moving toward the source. Thus, the top sign would be used in both the numerator (plus) and denominator (minus), and the general form of the Doppler effect equation would be:

$$f' = f\frac{(v + v_D)}{(v - v_S)}$$

8. Geometrical Optics

1. This is another application of Snell's law. Consider only the initial and final media because light enters and exits the pane of glass at the same angle:

$$n_{smoke} \sin \theta_{smoke} = n_{air} \sin \theta_{air}$$

$$\theta_{air} = \sin^{-1}\frac{n_{smoke} \sin \theta_{smoke}}{n_{air}}$$

$$= \sin^{-1}\frac{1.1 \times \sin 45°}{1} \approx \sin^{-1} 0.778$$

$$\theta_{air} \approx 51°$$

2. This question requires another two applications of Snell's law:

$$n_{air} \sin \theta_{air} = n_{diamond} \sin \theta_{diamond}$$

$$\theta_{diamond} = \sin^{-1} \frac{n_{air} \sin \theta_{air}}{n_{diamond}}$$

$$= \sin^{-1} \frac{1 \times \sin 35°}{2.4} \approx \sin^{-1} 0.239$$

$$\theta_{diamond} \approx 13.8°$$

and

$$n_{air} \sin \theta_{air} = n_{zircon} \sin \theta_{zircon}$$

$$\theta_{zircon} = \sin^{-1} \frac{n_{air} \sin \theta_{air}}{n_{zircon}}$$

$$= \sin^{-1} \frac{1 \times \sin 35°}{1.9} \approx \sin^{-1} 0.301$$

$$\theta_{zircon} \approx 17.6°$$

3. The critical angle is the angle at which light can no longer exit a medium, and instead is totally internally reflected. This implies that the refracted angle must be greater than or equal to 90°. The critical angle can be found as follows:

$$n_1 \sin \theta_c = n_2 \sin 90° \rightarrow \theta_c = \sin^{-1} \frac{n_2}{n_1}$$

We are given the relevant indices of refraction in the original question:

$$\theta_c = \sin^{-1} \frac{1}{1.34} \approx 48°$$

9. Nuclear Reactions

1. It is not possible for neptunium to decay into lead through a series of alpha decays. Each alpha decay reduces the atom's atomic number by two, but the difference between neptunium and lead's atomic numbers (93 – 82 = 11) is not divisible by two.

2. Balance the decay equation to determine how many β-particles are ejected:

$$^{214}_{84}\text{Po} \rightarrow {}^{214}_{86}\text{Rn} + 2\,{}^{0}_{-1}\beta$$

Two β-particles are ejected during this decay reaction.

3. Balance the decay equation to determine how many α-particles are ejected:

$$^{226}_{92}\text{U} \rightarrow {}^{218}_{86}\text{Rn} + 2\,{}^{0}_{+1}\beta + 2\,{}^{4}_{2}\alpha$$

Two α-particles are ejected during this decay reaction.

10. Dimensional Analysis

1. First, rearrange the universal law of gravitation to solve for G:

$$F_g = \frac{Gm_1m_2}{r^2} \rightarrow G = \frac{F_g r^2}{m_1m_2}$$

Now, plug in the relevant units:

$$[G] = \frac{N \cdot m^2}{kg \cdot kg} = \frac{\left(\frac{kg \cdot m}{s^2}\right)m^2}{kg^2} = \frac{m^3}{s^2 \cdot kg}$$

2. The stronger the electric field, the stronger the electrostatic force. Because the electrostatic force is the restoring force in this question, we would expect that the larger its magnitude, the faster the dipole oscillates. Hence, the period should *decrease* as a function of the electric field—not *increase*. This makes the given equation unreasonable.

3. First, rearrange an equation that uses the permeability of free space:

$$B = \frac{\mu_0 I}{2r} \rightarrow \mu_0 = \frac{2rB}{I}$$

The units for r, a distance, are meters. The units for magnetic field are teslas, where $1\ T = 1\frac{N}{m \cdot A}$. (If you did not remember this conversion factor, it can be determined from the equation $F = ILB \sin \theta$, where F is force in newtons, I is current in ampères, and L is length in meters.) Finally, the units for current are ampères. Now, plug in the relevant units:

$$[\mu_0] = \frac{[m]\left[\frac{N}{m \cdot A}\right]}{[A]} = \frac{N}{A^2} = \frac{\frac{kg \cdot m}{s^2}}{\left(\frac{C}{s}\right)^2} = \frac{kg \cdot m}{C^2}$$

11. Study Design

1. Correlation does not necessarily indicate causation. In this case, confounding likely explains the relationship between shark attacks and ice cream sales—namely, both of these variables are likely increased in summer months, when more individuals visit the beach, and when salt water climes may be more habitable for sharks. Ice cream sales also likely increase during summer months.

2. Respect for persons is maintained in any research context by providing the opportunity for informed consent, by allowing subjects to withdraw at any time, and by providing a debriefing at the end of the study period. Electroconvulsive therapy is still used in many psychiatric illnesses, including severe depression and suicidality, as well as psychotic disorders.

3. Causality in an experimental study is established through the use of at least one positive or negative control. Causality in an observational study cannot be proven, but can be supported through the use of Hill's criteria (temporality, strength, dose–response relationships, consistency, plausibility, consideration of alternative explanations, experiments, specificity, and coherence).

12. Graphical Analysis

1. Pie charts are generally poor at information dissemination with more than four categories, and for information that is not meant to add up to a whole. In addition, the colors have poor contrast and there is no written explanation or key. The chart is unlabeled, which leaves it uninterpretable.

2. For graphs, always start by checking the axes, including the relevant quantities, the units, and if there are any interruptions in the axes (indicated by wavy lines). Then, look for points of intersection, maxima, minima, differences in slope, or anything else noteworthy in the graph. For tables, always check for unusual values (zero, infinity, and so on) and make sure that the referenced categories are comparable.

3. These lines can be written in slope–intercept form. The slope of the red line is:

$$m_{red} = \frac{\Delta y}{\Delta x} = \frac{10 - 9}{-1 - 0} = \frac{1}{-1} = -1$$

This line's y-intercept is (0, 9), so the equation of the line is $y = -x + 9$.

The slope of the blue line is:

$$m_{blue} = \frac{\Delta y}{\Delta x} = \frac{0 - 1}{-0.33 - 0} = \frac{-1}{-0.33} = 3$$

This line's y-intercept is (0, 1), so the equation of the line is $y = 3x + 1$.

To determine their point of intersection, set the two equations equal to each other:

$$y = -x + 9$$
$$y = 3x + 1$$
$$-x + 9 = 3x + 1$$
$$4x = 8$$
$$x = 2$$
$$y = -x + 9 = -2 + 9 = 7$$

Thus, these lines intersect at the point (2, 7).